Related titles

PLATED STRUCTURES: STABILITY AND STRENGTH
edited by R. Narayanan

1. Longitudinally and Transversely Reinforced Plate Girders   H. R. EVANS
2. Ultimate Capacity of Plate Girders with Openings in Their Webs   R. NARAYANAN
3. Patch Loading on Plate Girders   T. M. ROBERTS
4. Optimum Rigidity of Stiffeners of Webs and Flanges   M. ŠKALOUD
5. Ultimate Capacity of Stiffened Plates in Compression   N. W. MURRAY
6. Shear Lag in Box Girders   V. KŘISTEK
7. Compressive Strength of Biaxially Loaded Plates   R. NARAYANAN and N. E. SHANMUGAM
8. The Interaction of Direct and Shear Stresses on Plate Panels   J. E. HARDING

SHELL STRUCTURES: STABILITY AND STRENGTH
edited by R. Narayanan

1. Stability and Collapse Analysis of Axially Compressed Cylindrical Shells   A. CHAJES
2. Stiffened Cylindrical Shells under Axial and Pressure Loading   J. G. A. CROLL
3. Ring Stiffened Cylinders under External Pressure   S. KENDRICK
4. Composite, Double-Skin Sandwich Pressure Vessels   P. MONTAGUE
5. Fabricated Tubular Columns used in Offshore Structures   W. F. CHEN and H. SUGIMOTO
6. Collision Damage and Residual Strength of Tubular Members in Steel Offshore Structures   T. H. SØREIDE and D. KAVLIE
7. Collapse Behaviour of Submarine Pipelines   P. E. DE WINTER, J. W. B. STARK and J. WITTEVEEN
8. Cold-Formed Steel Shells   G. ABDEL-SAYED
9. Torispherical Shells   G. D. GALLETLY
10. Tensegric Shells   O. VILNAY

STEEL FRAMED STRUCTURES: STABILITY AND STRENGTH
edited by R. Narayanan

1. Frame Instability and the Plastic Design of Rigid Frames   M. R. HORNE
2. Matrix Methods of Analysis of Multi-Storeyed Sway Frames   T. M. ROBERTS
3. Design of Multi-Storey Steel Frames to Sway Deflection Limitations   D. ANDERSON
4. Interbraced Columns and Beams   I. C. MEDLAND and C. M. SEGEDIN
5. Elastic Stability of Rigidly and Semi-Rigidly Connected Unbraced Frames   G. J. SIMITSES and A. S. VLAHINOS
6. Beam-to-Column Moment-Resisting Connections   W. F. CHEN and E. M. LUI
7. Flexibly Connected Steel Frames   K. H. GERSTLE
8. Portal Frames Composed of Cold-Formed Channel- and Z-Sections   G. J. HANCOCK
9. Braced Steel Arches   S. KOMATSU
10. Member Stability in Portal Frames   L. J. MORRIS and K. NAKANE

CONCRETE FRAMED STRUCTURES

Stability and Strength

Edited by

R. NARAYANAN

M.Sc. (Eng.), Ph.D., D.I.C., C.Eng., F.I.C.E., F.I.Struct.E., F.I.E.
Reader in Civil and Structural Engineering, University College, Cardiff, United Kingdom

ELSEVIER APPLIED SCIENCE PUBLISHERS
LONDON and NEW YORK

ELSEVIER APPLIED SCIENCE PUBLISHERS LTD
Crown House, Linton Road, Barking, Essex IG11 8JU, England

Sole Distributor in the USA and Canada
ELSEVIER SCIENCE PUBLISHING CO., INC.
52 Vanderbilt Avenue, New York, NY 10017, USA

WITH 11 TABLES AND 123 ILLUSTRATIONS

© ELSEVIER APPLIED SCIENCE PUBLISHERS LTD 1986

British Library Cataloguing in Publication Data

Concrete framed structures: stability and
  strength.
  1. Concrete constructions   2. Structural
  frames   3. Strength of materials
  I. Narayanan, R.
  624.1'8343   TA681.5

Library of Congress Cataloging-in-Publication Data

Concrete framed structures.

  Bibliography: p.
  Includes index.
    1. Reinforced concrete construction.   2. Structural
  frames.   3. Structural stability.   I. Narayanan, R.
  TA683.5.F8C66   1986   624.1'8341   86-19847

ISBN 1-85166-014-3

The selection and presentation of material and the opinions expressed in this publication are the sole responsibility of the authors concerned.

Special regulations for readers in the USA
This publication has been registered with the Copyright Clearance Center Inc. (CCC), Salem, Massachusetts. Information can be obtained from the CCC about conditions under which photocopies of parts of this publication may be made in the USA. All other copyright questions, including photocopying outside of the USA, should be referred to the publisher.

All rights reserved. No part of this publication may be reproduced, stored in a retrieval system, or transmitted in any form or by any means, electronic, mechanical, photocopying, recording, or otherwise, without the prior written permission of the publisher.

Phototypesetting by Tech-Set, Gateshead, Tyne & Wear.
Printed in Great Britain by Galliard (Printers) Ltd, Great Yarmouth.

PREFACE

It is my privilege to write a short preface to this book on concrete framed structures, the sixth in the planned set of volumes on stability and strength of structures.

Many excellent books by distinguished authors are available which describe the behaviour of concrete framed structures quite adequately; most of these are heavily code-oriented and serve the needs of the designer by providing an extensive range of design charts, tables and examples. On the other hand, the main purpose of this book is to bring about a basic understanding of the background to the material found in many of the modern codes and to describe developments currently taking place that are likely to influence future codes.

We have continued the policy of inviting several expert contributors to write a chapter each, so that the reader is presented with the 'state-of-the-art' on a number of related topics along with sufficient introductory material; the topics chosen are not intended to be exhaustive but they do reflect the diversity of problems within the field of concrete framed structures.

The book begins with an introduction to the stability of reinforced concrete building frames and thereafter proceeds to discuss recent developments in many facets of the subject, viz. compression members, optimum design, biaxial bending, torsion, deep beams, shear walls and flat slabs. The final chapter attempts to discuss the progressive collapse of structures, an aspect of structural behaviour which received little attention in the past but which has gained importance in view of the many catastrophic failures that have occurred recently.

As editor, I wish to express my gratitude to all the contributors for the willing cooperation they extended in producing this volume. I sincerely hope that readers find the book informative and stimulating.

R. NARAYANAN

CONTENTS

Preface v

List of Contributors ix

1. Stability of Reinforced Concrete Building Frames . . . 1
   J. G. MACGREGOR

2. Stability of Compression Members 43
   U. QUAST

3. Nonlinear Analysis and Optimal Design of Concrete Framed Structures 71
   C. S. KRISHNAMOORTHY

4. Reinforced Concrete Columns in Biaxial Bending . . 111
   I. E. HARIK and H. GESUND

5. Design of Concrete Structures for Torsion 133
   R. NARAYANAN

6. Reinforced Concrete Deep Beams 169
   F. K. KONG

7. Column-supported Shear Walls 193
   L. CERNY and R. LEON

8. Design of Reinforced Concrete Flat Slabs 217
   P. E. REGAN

9. Progressive Collapse of Slab Structures 249
   D. MITCHELL and W. D. COOK

Index 279

LIST OF CONTRIBUTORS

L. Cerny
Department of Civil and Mineral Engineering, University of Minnesota, Minneapolis 55455-0220, Minnesota, USA

W. D. Cook
Researcher, Department of Civil Engineering and Applied Mechanics, McGill University, Montreal H3A 2K6, Canada

H. Gesund
Department of Civil Engineering, University of Kentucky, Lexington, Kentucky 40506-0046, USA

I. E. Harik
Department of Civil Engineering, University of Kentucky, Lexington, Kentucky 40506-0046, USA

F. K. Kong
Professor of Structural Engineering, University of Newcastle upon Tyne, Cassie Building, Claremont Road, Newcastle upon Tyne NE1 7RU, UK

C. S. Krishnamoorthy
Professor of Structural Engineering, Department of Civil Engineering, Indian Institute of Technology, Madras 600 036, India

R. Leon
Department of Civil and Mineral Engineering, University of Minnesota, Minneapolis 55455-0220, Minnesota, USA

J. G. MacGregor
University Professor of Civil Engineering, University of Alberta, Edmonton T6G 2G7, Canada

D. MITCHELL
Associate Professor, Department of Civil Engineering and Applied Mechanics, McGill University, 817 Sherbrooke Street West, Montreal H3A 2K6, Canada

R. NARAYANAN
Reader in Civil and Structural Engineering, University College, Newport Road, Cardiff CF2 1TA, UK

U. QUAST
Professor of Concrete Structures, Arbeitsbereich Massivbau, Technische Universität Hamburg-Harburg, Lauenbruch-Ost 1, Postfach 90 14 03, 2100 Hamburg 90, Federal Republic of Germany

P. E. REGAN
Reader in Civil Engineering, Polytechnic of Central London, 35 Marylebone Road, London NW1 5LS, UK

Chapter 1

STABILITY OF REINFORCED CONCRETE BUILDING FRAMES

J. G. MacGregor

Department of Civil Engineering, University of Alberta, Edmonton, Canada

SUMMARY

The current design procedures for slender concrete columns are based on elastic stability theory applied to a single-curvature, hinged-end column. Approximations were introduced to extend this to columns with other moment diagrams, to reinforced concrete columns that crack and creep, and finally to columns in frames. The design procedures from a number of countries are discussed and it is concluded that the most questionable approximations involve the choice of the EI of the cross-section, the treatment of creep, and the use of elastic effective length factors to approximate frame behavior. In recent years a number of concepts have been developed to improve the treatment of columns in frames. Several of these are discussed. The final portion of the chapter discusses the stability of three-dimensional frames.

NOTATION

C_m Equivalent moment factor, eqn (1.12)
d Distance from extreme compression fiber to steel on opposite face of column
e Eccentricity of load at end of column
e_c Eccentricity introduced to account for creep deflections
e_g First-order eccentricity of the sustained loads
e_{01} and e_{02} Smaller and larger first-order end eccentricities
e_2 Second-order eccentricity due to column deflections

E_c	Modulus of elasticity of concrete
E_s	Modulus of elasticity of steel
f'_c	Specified strength of concrete
h	Overall thickness of column section
I_g	Moment of inertia of gross section
I_s	Moment of inertia of reinforcement about centroid of gross section
k	Effective length factor
K_b	Flexural stiffness of beam
K_c	Flexural stiffness of column
L	Column length, storey height
L_u	Unsupported height of column
M	Moment
M_b	Moment at balanced failure condition
M_D	Sustained load moment
M_f	Moment corresponding to a strain distribution with $\varepsilon_s = \pm \varepsilon_y$ in the steel closest to the two faces of the column
M_L	Moment due to short time loads
M_{ns}	Moment due to loads which do not cause appreciable sway
M_s	Moment due to loads which cause appreciable sway
M_0	First-order moments
M_1	Smaller end moment on column, positive if column is bent in single curvature
M_2	Larger end moment on column, always positive
N	Axial load
N_b	Axial load at balanced failure condition
N_c	Critical load $= \pi^2 EI/(kL)^2$
N_E	Critical load of a hinged-end column $= \pi^2 EI/L^2$
N_g	Axial load due to sustained loads
Q	Stability coefficient, eqn (1.33)
r	Radius of gyration
r_j	Radial distance from center of rotation to the jth column
T	Total magnified torque
T_1	First-order torque
V	Shear in a storey
W_i	Total gravity load in the ith storey
α	Ratio of sustained axial loads to total axial loads
β	Ratio of sustained load moment to total load moment
β_d	Absolute value of the ratio of the factored dead load moment to the factored total load moment

γ	Flexibility factor
δ	Magnifier
δ_D	Deflection magnifier
δ_s	Sway deflection or moment magnifier
Δ	Total magnified deflection
Δ_0	First-order deflection
ε_y	Yield strain of reinforcement
ρ_f	Radius of curvature corresponding to M_f
ρ_t	Ratio of total area of reinforcement to the area of the gross cross-section of the column
$1/\rho$	Curvature
ϕ	Creep coefficient
ϕ_0	First-order rotation of building about center of rotation
ΣN	Sum of axial forces in all the columns in a storey
ΣV	Sum of shears in all the columns in a storey

1.1 INTRODUCTION

This chapter discusses present and future methods of design of reinforced concrete columns. It consists of four parts. First the response of an idealized hinged-end column is reviewed briefly, followed by similar reviews of the behavior of concrete columns in braced frames and sway frames. In Section 1.4, the derivation of the present design methods for concrete columns is presented in a step-by-step fashion. It is seen that there are a number of areas where improvements are warranted. Section 1.5 describes four new developments in column design procedures. Although some of these are recognized in the newest design codes, there is much scope for development. Sections 1.6 and 1.7 discuss the elastic stability of planar structures and three-dimensional structures.

1.2 BEHAVIOR OF HINGED, REINFORCED CONCRETE COLUMNS

1.2.1 Moment Magnification

When the slender hinged column shown in Fig. 1.1 is loaded, it displaces laterally and at mid-height it must resist an axial load N and a moment $N(e + \Delta)$. The effect of the additional moment, $N\Delta$, can be seen in Fig. 1.2. The line O–A represents the relationship between the

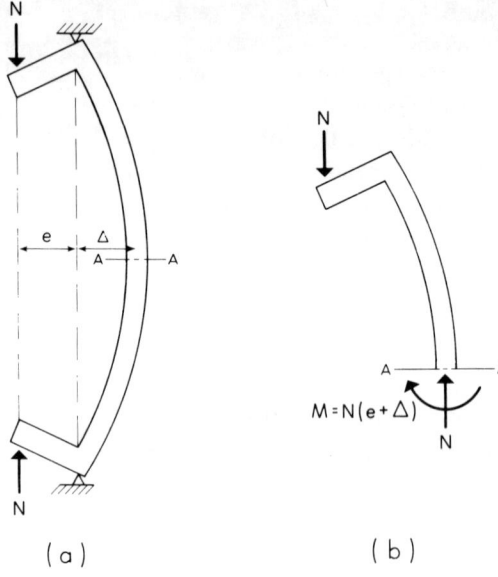

FIG. 1.1. Slender column under load. Reproduced from MacGregor *et al.* (1970) by permission of the American Concrete Institute.

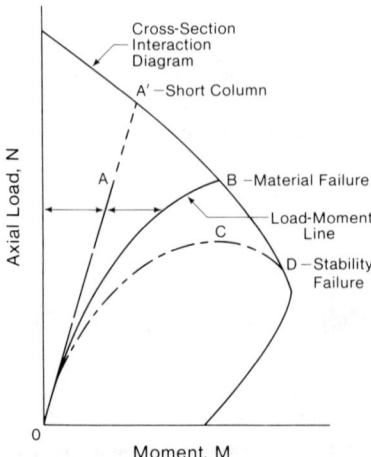

FIG. 1.2. Loads and moments at mid-height for two slender columns.

load and the moment at the end of the column as the load N is increased. At any load N, the moment at the end of the column is Ne. The curved line O–B represents the load and moment at mid-height of the column. At any load N, this is made up of the end moment Ne plus the moment $N\Delta$. This column fails when the 'load–moment line' O–B intersects the interaction diagram for the column at point B. If the column had not deflected it would fail at the load corresponding to point A'. The reduction in axial load capacity from A' to B is referred to as the 'slenderness effect'.

A very slender column might display the behavior shown by the load–moment line O–C–D in Fig. 1.2. Here the column becomes unstable when the load and mid-height moment reach point C. Further deflection occurs without increase in applied load. This column is said to develop a 'stability' failure as contrasted to the shorter column which developed a 'material failure' when the load and maximum moment at the mid-height of the column corresponded to a point on the interaction curve for the cross-section.

1.2.2 Unequal End Moments

In Fig. 1.1 the end eccentricities were equal, with the result that e and Δ were directly additive at all points along the column. If the end moments are not equal, as shown in Fig. 1.3, the maximum e and the maximum Δ occur at different points along the column. In such a case, the maximum

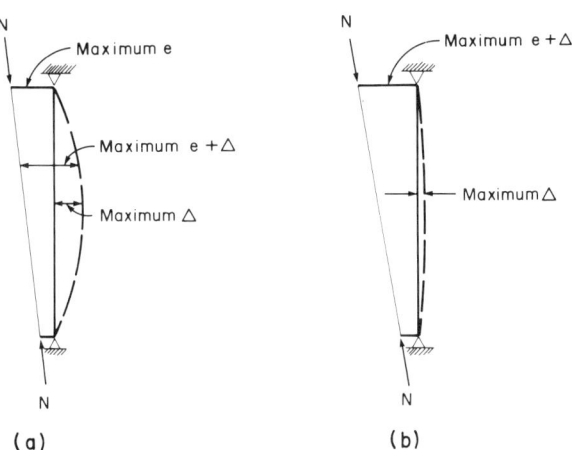

FIG. 1.3. Columns with unequal end moments.

e and the maximum Δ cannot be added directly. Two cases can be identified. In Fig. 1.3(a) the end eccentricities are small and, because the column is quite slender, the deflections are large. Here the maximum sum of $e + \Delta$ occurs between the ends of the column. Since $e + \Delta$ at this point exceeds e_2, the column is weakened by the slenderness effect. In Fig. 1.3(b) the maximum sum of $e + \Delta$ occurs at one end of the column. This column is not weakened by slenderness.

1.2.3 Effect of Sustained Loads

Figure 1.4 shows the load–moment histories of two columns loaded quickly to loads N_1 and N_2 which were then held constant for a long period of time. During this time the deflections increased, and hence the $N\Delta$ moments increased. Column 1 failed during the sustained loading period. This is referred to as 'creep buckling' and occurs only for very heavily loaded columns. This type of behavior is discussed by Gvozdev and Chistiakov (1972).

Column 2 did not fail under sustained loads and an additional rapidly applied load was required to cause failure. The column was weakened by the sustained loads because the creep deflections increased the overall moment at the failure section. Green and Breen (1984) present column tests displaying both types of column behavior.

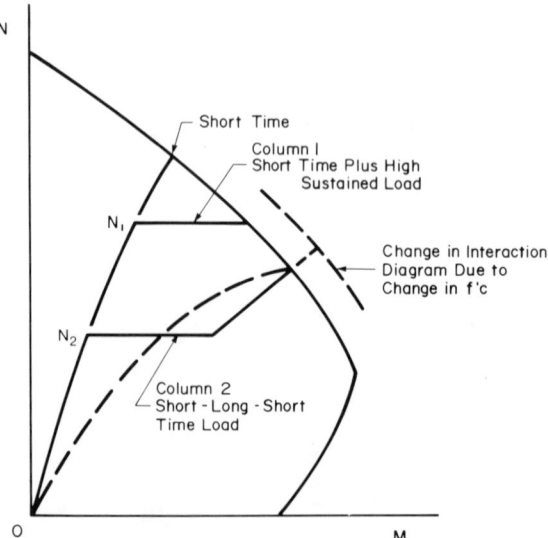

FIG. 1.4. Effect of sustained loads on moments in columns.

1.3 BEHAVIOR OF COLUMNS IN FRAMES

1.3.1 Behavior of Columns in Braced Frames

Figure 1.5 shows a simple braced frame. The applied load moment Ne is divided between the column and beam in the ratio of their stiffnesses. Thus

$$M_c = Ne\left(\frac{K_c}{K_c + K_b}\right) \tag{1.1}$$

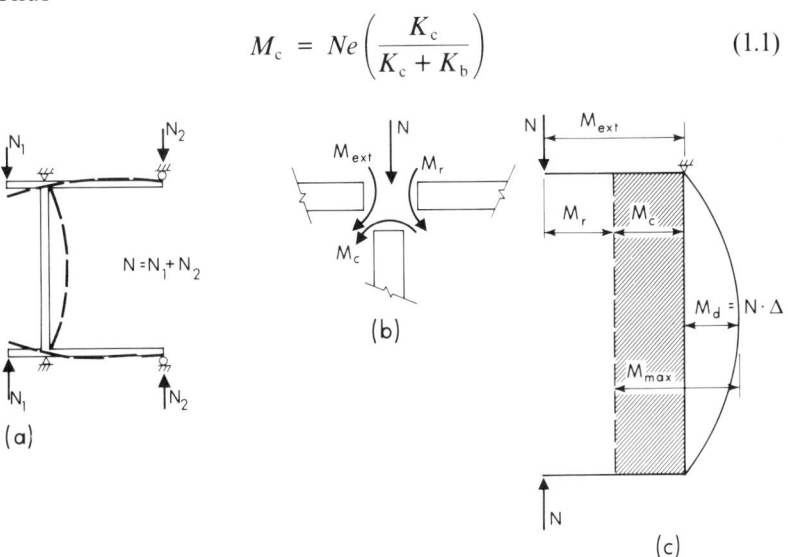

FIG. 1.5. Moments in a restrained column: (a) model of restrained column; (b) distribution of moments at joint; (c) moments in the column. Reproduced from MacGregor *et al.* (1970) by permission of the American Concrete Institute.

As the load N is increased, two things happen simultaneously. First, the deflection Δ increases, causing an increase in the $N\Delta$ moment. Secondly, the column stiffness K_c decreases. As a result of the decrease in column stiffness, the end moment M_c decreases, offsetting some of the increase in the $N\Delta$ moment. This is illustrated in Fig. 1.6 which shows the measured load–moment curves for two points in a column in a symmetrically loaded frame tested by Furlong and Ferguson (1966). At the end of the column (Section B) the moment decreased as the axial load was increased. At mid-height (Section A) the moment increased because the $N\Delta$ moments more than offset the reduction in the end moments.

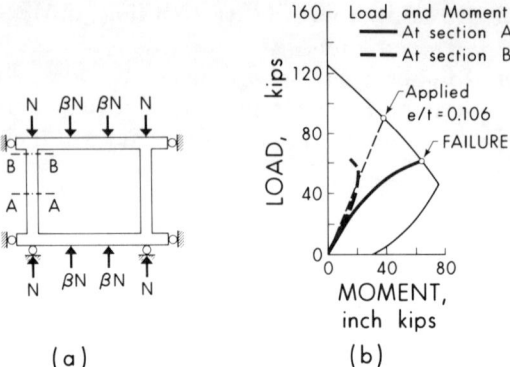

FIG. 1.6. Load–moment history in a braced frame: (a) test specimen; (b) measured moment response. Reproduced from MacGregor *et al.* (1970) by permission of the American Concrete Institute.

With increasing axial load the end moments on an elastically restrained column reverse sign and the final deflected shape approaches the elastic buckling mode of an axially loaded column as shown in Fig. 1.7(a). For columns loaded in reversed curvature the moment at one end changes sign as shown in Figs. 1.7(b) and (c). For practical columns it is likely that the column will develop a material failure before it reaches the elastic mode shape. Cranston (1972) presents an excellent discussion of these effects, tracing loads, curvatures, deflections and moments in a series of elastically restrained concrete columns as they are loaded to failure.

Creep accentuates both the reduction in end moment and the increase in deflections. Manuel and MacGregor (1967) found that for columns with L_u/h up to about 12 ($L_u/r \simeq 35$) in braced frames, the reduction in end moments due to creep was larger than the $N\Delta$ moments, resulting in an increase in column capacity. This possibility is not considered in the current design codes. For longer columns there was a reduction in capacity under sustained loads.

1.3.2 Behavior of Columns in Unbraced Frames

When lateral loads act on an unbraced frame it deflects laterally as shown in Fig. 1.8. The summation of the moments at the ends of the columns must balance the moments due to the lateral displacement of the structure. Thus

$$\Sigma M_\text{T} + \Sigma M_\text{B} = VL + \Sigma N\Delta \qquad (1.2)$$

where M_T and M_B are the moments at the tops and bottoms of the columns. The $\Sigma N\Delta$ term reflects the effect of vertical loads on the equilibrium and will be referred to later in this chapter as the 'lateral drift effect'.

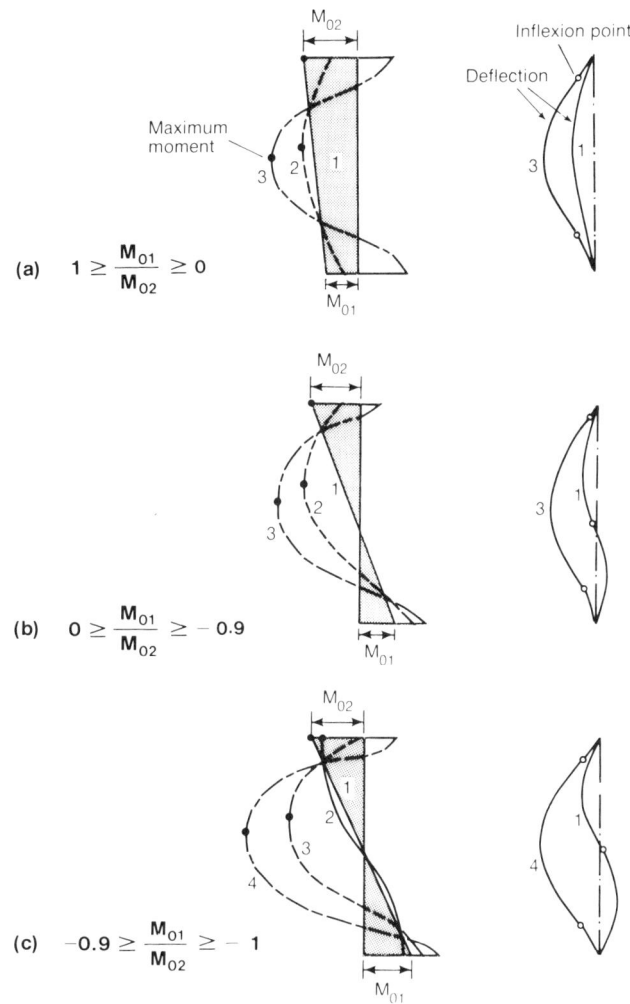

FIG. 1.7. Bending moments and deflected shapes of symmetrically restrained columns under increasing compression. Reproduced from Lai et al. (1983) by permission of the American Society of Civil Engineers.

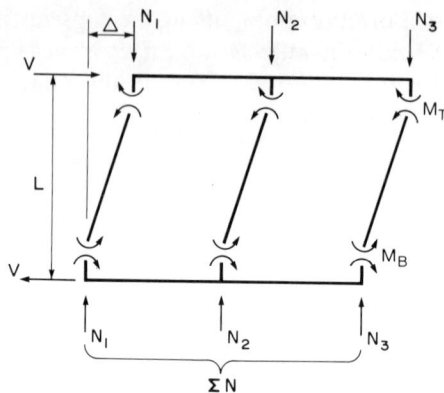

FIG. 1.8. Equilibrium of a sway frame.

Several things are important here. First, the maximum lateral load moments and the maximum $N\Delta$ moments both occur at the ends of the columns and hence are directly additive. Secondly, the upper floor moves as a unit relative to the lower floor so that the value of Δ is the same, or nearly so, for all columns. As a result, ΣN is more important in the analysis than the value of N on a given column. Thirdly, a reduction in the column stiffness due to axial loads and cracking leads to an increase in deflections and hence an increase in the $\Sigma N\Delta$ moments. This differs from the braced column case shown in Fig. 1.6 where the softening of the column causes a reduction in the rate of increase of moments at the ends of the columns. Finally, the beams must be able to withstand the increased end moments in the columns. If they are not able to, hinges will form in the beams forming a mechanism.

1.4 DESIGN OF SLENDER REINFORCED CONCRETE COLUMNS — 1950 TO 1975

The design procedures for slender reinforced concrete columns in most modern codes were originally derived for the idealized case of an elastic, hinged-end column bent in symmetrical single curvature. The solution adopted for this idealized case was then empirically extended:

(a) from the case of symmetrical single curvature to the case of unequal end moments;

STABILITY OF REINFORCED CONCRETE BUILDING FRAMES   11

(b) from the case of 'elastic' columns to reinforced concrete columns which have non-linear moment curvature response and which undergo creep;
(c) from the basic case of hinged columns to apply to columns in frames.

While the basic analysis of the elastic, hinged-end column bent in single curvature is a close approximation to the truth, each of the additional extensions reduces the accuracy and credibility of the final design method.

1.4.1 Moment Magnifier for Elastic, Hinged-End, Single-Curvature Columns

As the load N on the column in Fig. 1.1 approaches the Euler buckling load, the deflection Δ increases, approaching infinity. The total deflection, Δ, is closely approximated by

$$\Delta = \Delta_1 \delta_D \tag{1.3}$$

where Δ_1 is the first-order deflection due to the end moments Ne and δ_D is a deflection magnifier:

$$\delta_D = \frac{1}{1 - N/N_E} \tag{1.4}$$

where N_E is the Euler buckling load of a hinged-end column, $N_E = \pi^2 EI/L^2$. For N/N_E less than 0·7, eqn (1.3) is accurate within 2% for an elastic column similar to Fig. 1.1.

The maximum moment in the column in Fig. 1.1 is

$$M = Ne + N\Delta_1 \left(\frac{1}{1 - N/N_E}\right) \tag{1.5}$$

Substituting the equation for Δ_1 into eqn (1.5) and rearranging gives

$$M = Ne \left(\frac{1 + 0.234 N/N_E}{1 - N/N_E}\right) \tag{1.6}$$

For an elastic column similar to Fig. 1.1 the moment magnifier given by eqn (1.6) will give moments M within $\pm$ 0·2% of the exact elastic solution for $N \leqslant 0.5N_E$ and within 0 to -2% for $N > 0.5N_E$. The coefficient 0·234 only applies to a hinged-end column deflected in symmetrical single curvature by equal end moments as shown in Fig. 1.1. For other loading conditions it varies from $+0.25$ to -0.4.

1.4.1.1 *Swiss Code Procedure*

In the 1976 revisions to the Swiss Code (SIA, 1976), eqn (1.5) is used to design columns. Since the calculation of Δ_1 takes into account the actual distribution of first-order moments along the length of the column, the problem of variations in the numerical coefficients in eqn (1.6) is avoided.

1.4.1.2 *ACI Code Procedure*

Recognizing the variations in the numerical coefficients in eqn (1.6), the ACI (1983) Code sets the coefficient equal to zero:

$$M = (Ne)\delta \quad (1.7)$$

where

$$\delta = \left(\frac{1}{1 - N/N_E}\right) \quad (1.8)$$

This, in effect, amounts to replacing the moment magnifier in eqn (1.6) with δ_D. For a hinged-end, elastic column similar to Fig. 1.1, eqns (1.7) and (1.8) underestimate M by up to 12% at $N/N_E = 0.5$ and by lesser amounts for columns with unequal end moments.

1.4.1.3 *CEB Second-order Eccentricity Method*

In the 1978 CEB Model Code, three design methods were presented (CEB, 1978a): a so-called general method based on calculation of the deflected shape of the column from the moment–curvature diagram of the cross-section; a method in which a second-order eccentricity is added to the first-order eccentricity; and the model column method which is a special case of the other two methods.

For a hinged column bent in single curvature the second-order eccentricity method is based on the total eccentricity $(e + \Delta)$ or e_{tot}:

$$e_{tot} = e + e_a + e_2 \quad (1.9)$$

where e is the first-order end eccentricity, e_a is an empirical additional eccentricity to account for geometric inaccuracies, and e_2 is the second-order eccentricity due to the column deflection Δ. This is given by

$$e_2 = \frac{L^2}{10}\frac{1}{\rho} \quad (1.10)$$

where L is the effective length of the column and $1/\rho$ is the curvature at the critical section. If a hinged column was bent in a circular arc, the coefficient 10 in eqn (1.10) would be 8; if bent in a sine-wave, it would be $\pi^2 = 9.87$.

STABILITY OF REINFORCED CONCRETE BUILDING FRAMES 13

The major problem is choosing which curvature value should be used in eqn (1.10). This will be discussed more fully later.

Disregarding e_a, substituting $N_E = 10EI/L^2$ and rearranging eqn (1.9) gives the ACI moment-magnifier equation (1.7) and (1.8). The major differences between the two codes is that the ACI (1983) Code specifies a value of EI for use in calculating N_E while the CEB (1978a) Model Code does not. The British Code CP110 is based on the CEB second-order eccentricity method (BSI, 1972).

1.4.2 Correction for Unequal End Moments

Section 1.2.2 and Fig. 1.3 illustrate the effects of unequal end moments on the strength of columns. The ACI (1983) Code accounts for this by replacing the columns in Fig. 1.3 with an equivalent column bent in single curvature with equal end moments $C_m M_2$ where C_m is selected so that the magnified moment in the equivalent column gives a reasonable approximation of the true magnified moment of the column with unequal end moments, where

$$M = \delta C_m M_2 \geq M_2 \qquad (1.11)$$

In eqn (1.11) M_2 is the larger end moment on the column.

If δC_m exceeds 1·0, the maximum moment occurs between the ends of the column (Fig. 1.3a), and if δC_m is less than or equal to 1·0, the maximum moment occurs at the end of the column (Fig. 1.3b). A reasonable approximation of C_m is

$$C_m = 0·6 + 0·4 M_1/M_2 \geq 0·4 \qquad (1.12)$$

The ratio M_1/M_2 is positive when the column is bent in single curvature. For sway frames C_m is always equal to 1·0 since the lateral load moments and $N\Delta$ moments both occur at the ends of the column and hence are directly additive.

The CEB (1978a) Model Code accounts for unequal end eccentricities by modifying them using an equation derived from eqn (1.12).

1.4.3 Modification of Elastically Derived Moment Magnifiers to Apply to Reinforced Concrete Columns

Up to this point the derivation has been restricted to elastic columns with constant EI. For reinforced concrete columns EI is not constant (a) because of inelastic action of the concrete and reinforcement, and the cracking of the concrete, and (b) because of creep strains in the concrete. Current design approximations for reinforced concrete columns were developed for short time loading and modified for creep.

1.4.3.1 *EI Expressions for Short Time Loads*

Figure 1.9 shows a moment–curvature diagram for a column cross-section subjected to a constant axial load. A radial line in this diagram has the slope

$$\frac{M}{(1/\rho)} = EI \qquad (1.13)$$

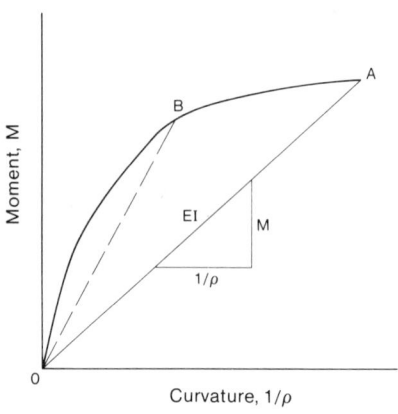

FIG. 1.9. Moment–curvature diagram for column cross-section.

The problem is to decide which radial line should be chosen. For columns developing material failures (see Fig. 1.2) the radial line O–A through the end of the moment–curvature diagram should be chosen. On the other hand, columns developing stability failures become unstable before the cross-section fails. Thus the critical value of EI will correspond to a steeper line such as O–B in Fig. 1.9. This is illustrated in Fig. 1.10 for a column bent in symmetrical single curvature.

Since Fig. 1.9 is plotted for a given value of axial force N, it can be converted to a diagram of the eccentricity e_i corresponding to each curvature, where $e_i = M/N$. This is shown by the curved line in Fig. 1.10. Superimposed on this curve is a sloping line with height $e + e_2$ where e is the end eccentricity of the column shown in Fig. 1.1 and e_2 is the deflection of a column with a given length L. As shown by eqn (1.10), e_2 increases linearly with curvature, $1/\rho$.

The intersection of the sloping line and the curved line at A in Fig. 1.10 represents the state of equilibrium reached by a column loaded

STABILITY OF REINFORCED CONCRETE BUILDING FRAMES

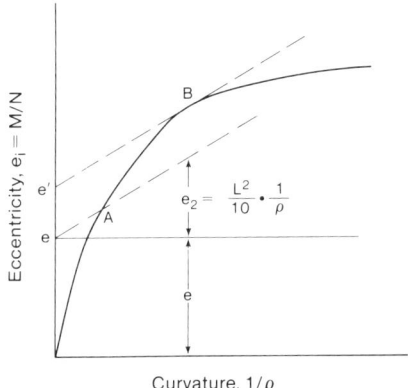

FIG. 1.10. Eccentricity–curvature diagram for column and eccentricity at midheight of the column.

with end eccentricity e, in this case a state of stable equilibrium. The total $e + e_2$ at mid-height is shown by the lower dashed line.

If the end eccentricity e is increased to e', the sloping line $e' + e_2$ representing the midspan eccentricity comes tangent to the curved eccentricity–curvature line at point B. This represents a state of unstable equilibrium. The corresponding moment and curvature are given by point B in Fig. 1.9 and the appropriate EI would be the slope of the radial line O–B in Fig. 1.9.

Various national codes have implemented these concepts in various ways. These fall into the three families discussed below.

(a) CEB Model Column procedure. The CEB (1978a) Model Code presents a so-called 'model column' method. The model column is equivalent to half of the column shown in Fig. 1.1 and is assumed fixed at mid-height and free to deflect and rotate at the top. The solution is based on eqns (1.9) and (1.10) and an eccentricity–curvature diagram or a moment–curvature diagram (CEB, 1978b). The solution is as shown in Fig. 1.10. It is not necessary, therefore, to have a value for EI or a value of $1/\rho$ at failure. It is necessary, however, to have a moment–curvature diagram for the column cross-section.

(b) Methods based on the EI or curvature corresponding to material failures. The British Code CP110 (BSI, 1972) is based on eqns (1.9) and

(1.10) but bases $1/\rho$ on an assumed curvature at failure. This was obtained by assuming the strain distribution in Fig. 1.11(a) with a concrete strain at crushing of $0.003\beta_c$, where β_c allows for creep, and a steel strain of ε_y:

$$\frac{1}{\rho_u} = \frac{(0.003\beta_c + \varepsilon_y)}{h} \quad (1.14)$$

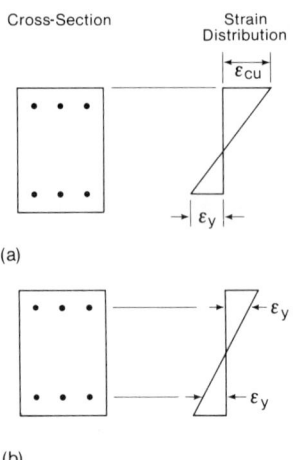

FIG. 1.11. Strain distributions used to define EI.

For simplicity in use this was written in terms of h, the overall depth, rather than d, the effective depth. To account for the fact that very long columns develop stability failures at smaller curvatures than this, and to allow for variations in $1/\rho_u$ as the ratio of axial load to balanced load N/N_b changed, this was rewritten as

$$\frac{1}{\rho_u} = \frac{\left(0.003\beta_c + \varepsilon_y - \dfrac{kL_u}{50\,000\,h}\right)}{h} K_1 \quad (1.15)$$

where kL_u is the effective length and K_1 is a factor equal to or less than 1.0 to account for the axial load ratio N/N_b. Substituting $\varepsilon_y = 0.002$ and $\beta_c = 1.25$ into eqn (1.15) gives

$$\frac{1}{\rho_u} = \frac{1}{175h}\left(1 - 0.0035\frac{kL_u}{h}\right) K_1 \quad (1.16)$$

This is used with eqns (1.9) and (1.10) to get the design moment. Because of the complexity of defining K_1, the designer is given the option of taking it equal to 1·0. Equation (1.16) is developed and compared to tests in the paper by Cranston (1972).

Menn (1974) proposed that EI be defined based on the moment M_f and curvature $(1/\rho)_f$ corresponding to the strain distribution shown in Fig. 1.11(b). Menn showed that the actual values of $EI = M/(1/\rho)$ corresponding to material failures were generally larger than this value except for tension failures for very low axial loads and columns with small eccentricities at the failure section. This is illustrated in Fig. 1.12. The height of the 'cliff' along the edge of the surface is almost constant over much of its length at a height greater than the value of EI corresponding to M_f and $(1/\rho)_f$. For axial loads less than the value N_f corresponding to the strain distribution in Fig. 1.11(b) it is necessary to reduce EI as a function of axial load. Contour maps of the surface shown in Fig. 1.12 have been published by Wood and Shaw (1979) and Argent and Banut (1979).

The Draft Australian Code (ASA, 1984) bases EI on the curvature at a balanced failure (Fig. 1.11a):

$$EI = \frac{M_b}{(1/\rho)} \quad (1.17)$$

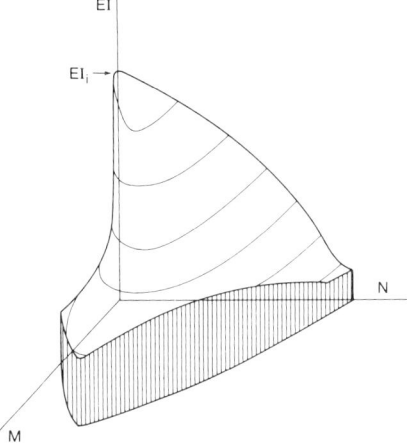

FIG. 1.12. Variation of EI as a function of load and moment. Reproduced from Menn (1974) by permission of the International Association for Bridge and Structural Engineering.

where M_b is the moment resistance of a section loaded with the limiting compressive strain ε_{cu} at one face and a tensile strain equal to the yield strain ε_y at the centroid of the tension steel; $1/\rho$ is the corresponding curvature. Assuming that $\varepsilon_{cu} = 0\cdot003$ and $\varepsilon_y = 0\cdot002$ gives

$$EI = 200\, d\, M_b \qquad (1.18)$$

Smith and Bridge (1984) show this to be a conservative estimate of EI. The conservatism increases as the slenderness of the columns increase, becoming very conservative for kL_u/r greater than 60.

The *FIP Recommendations on Practical Design* give a similar equation with 180 instead of 200 (FIP, 1982).

(c) Empirically derived EI values. For short time loads the ACI (1983) Code gives EI as

$$EI = 0\cdot2 E_c I_g + E_s I_s \qquad (1.19)$$

This was empirically derived by MacGregor et al. (1970) to correspond to EI values back-calculated from test results, and to approximate the knee in the moment–curvature diagrams in those cases where test data were not available.

Based on a large number of computer simulations, MacGregor et al. (1975) proposed that eqn (1.19) be rewritten

$$EI = E_c I_g (0\cdot2 + 1\cdot2 \rho_t E_s / E_c) \qquad (1.20)$$

where ρ_t is the ratio of the total area of steel to the area of the concrete.

1.4.3.2 *Effect of Sustained Loads*

Figure 1.4 illustrates two ways in which creep deflections may affect the strength of a slender column. In the ACI (1983) Code both these loading cases have been approximated by using a reduced EI to allow for creep strains. For column 2 in Fig. 1.4, this, in effect, replaces the actual load moment history with that shown by the dashed line. The reduced EI values are obtained by dividing eqn (1.19) by $(1 + \beta_d)$ where β_d is a creep factor defined as the absolute value of the ratio of the factored dead load moments M_D to the sum of the factored dead and live load moments $(M_D + M_L)$. This is expressed in terms of moments to account for the fact that dead loads are frequently applied at small eccentricities and

STABILITY OF REINFORCED CONCRETE BUILDING FRAMES   19

hence lead to small creep deflections while moments frequently result from rapidly applied loads such as wind or live loads. The EI of both the steel and the concrete are reduced since, with small amounts of steel, creep may lead to premature yielding of the reinforcement.

In the CEB (1978a) Model Code, creep is accounted for by adding a creep eccentricity e_c to eqn (1.9), or by multiplying the strain axis of the concrete stress–strain diagram used to calculate the moment–curvature diagram by $(1 + \alpha\beta\phi)$ where α is the ratio of sustained to total axial loads, and β is the corresponding ratio of moments. Recent analytical studies suggest the second procedure underestimates the effect of creep.

The creep eccentricity procedure was derived by Warner and Kordina (1975) and Warner et al. (1976) using Dischinger's creep theory. The calculation of the creep eccentricity is based on a hypothetical loading sequence in which the column is loaded quickly to its sustained load level N_g. This load is then assumed to remain on the column for a very long time, after which it is removed. The residual deflection after unloading is e_c, given by

$$e_c = e_g \exp\left(\frac{\phi N_g}{N_E - N_g}\right) - 1 \qquad (1.21)$$

where e_g is the first-order eccentricity of the sustained loads, ϕ is the creep coefficient for the period of sustained load and N_E is the critical load of the column.

In the Swiss Code (SIA, 1976), the concrete strains used to compute M_f are multiplied by $(1 + \alpha\phi)$. The British Code (BSI, 1972) simply assumes that the crushing strain is 1·25 times the short time crushing strain, i.e. $\beta_c = 1\cdot 25$ in eqns (1.14) and (1.15).

1.4.4 Modification of Hinged Column Magnifiers to Apply to Columns in Frames

The complex effect of end restraints described in Section 1.3 is approximated in all modern codes by empirically replacing the restrained column with an equivalent hinged-end column with a length kL_u, taken equal to the elastic effective length for a restrained axially loaded column. This implies that the restrained column can be replaced with a hinged-end column of length kL as shown in Fig. 1.13(a). This hypothetical hinged column is then loaded with the axial load and the first-order moments which acted at the ends of the restrained column.

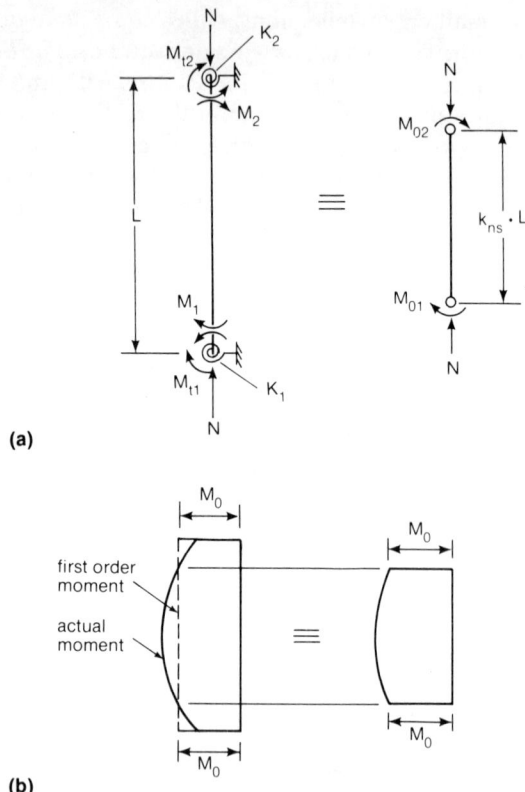

FIG. 1.13. Relationship between effective length and intersection length. Reproduced from Lai *et al.* (1983) by permission of the American Society of Civil Engineers.

In the ACI Code, design is based on eqns (1.7) and (1.8). Equation (1.8) is rewritten as

$$\delta = \frac{1}{1 - N/N_c} \tag{1.22}$$

where $N_c = \pi^2 EI/(kL_u)^2$.

Figure 1.13(b) compares the actual and assumed moment diagrams in a symmetrically loaded restrained column at an advanced stage of loading. The actual moment diagram intersects the first-order moment diagram at two points where the actual moments are identical to the first-order end moments. The column can then be considered as a pin-ended column subjected to end moments equal to the first-order end moments of the real restrained column, with a column length equal to the distance between the two intersection points. This distance is assumed to be equal to the effective length.

The validity of this assumption depends on the relative degree of softening of the column and end restraints as loading progresses. Lai et al. (1983) showed that for elastically restrained elastic columns kL_u underestimates the true effective length. On the other hand, for the elastically restrained reinforced concrete columns studied by Cranston (1972), kL can be shown to overestimate the intersection length. The behavior of a reinforced concrete column restrained by reinforced concrete beams should be intermediate.

For sway frames, one floor displaces relative to another as shown in Fig. 1.8. Thus, the magnification of sway moments is a function of $\Sigma N / \Sigma N_c$ for all columns in the storey where N_c is based on k greater than 1. This is accounted for in the ACI (1983) Code by rewriting eqn (1.22) as

$$\delta_s = \frac{1}{1 - \frac{\Sigma N}{\Sigma N_c}} \quad (1.23)$$

The British Code uses a similar concept (BSI, 1972): the value of (kL/h) averaged over all the columns in a storey is used to compute N_c for each column.

Hellesland (1976) has shown that for storeys or structures having columns of different lengths eqn (1.23) becomes

$$\delta_s = \frac{1}{1 - \frac{\Sigma(N/L)}{\Sigma(N_c/L)}} \quad (1.24)$$

1.4.5 Slenderness Bounds

Many columns are so stocky that it is not necessary to consider slenderness effects. Various codes specify the boundary beyond which slenderness must be considered in various ways.

The ACI (1983) Code regards columns in braced frames as short columns if

$$\frac{kL}{r} < 34 - 12 M_1/M_2 \quad (1.25)$$

where M_1/M_2 is the ratio of the first-order end moments and is positive for columns in single curvature. Slenderness must be considered in the design of columns in a sway frame if kL/r is greater than 22. This corresponds to the limit from eqn (1.25) for symmetrical single curvature bending because the maximum applied load moments and maximum second-order moments are directly additive in sway frames as they are in the single curvature case. Equation (1.25) was derived by back-calculation with the ACI Code design procedure (eqns 1.7, 1.12, 1.19 and 1.22) to get the slenderness ratio corresponding to a 5% reduction in strength.

The 1978 CEB Model Code regards columns as short if kL/r is less than 25. Slenderness effects must be considered but creep effects can be ignored if kL/r is between 25 and 50 and both slenderness and creep must be considered if kL/r exceeds 50. In the Commentary to a later section it is stated that the effect of unequal end eccentricities can be ignored if

$$\frac{kL}{r} < 50 - 25\, e_1/e_2 \qquad (1.26)$$

The CEB slenderness limits are assumed to apply equally to sway frames and braced frames.

More recent European studies (Menegotto, 1983) based on non-linear frame analyses allowing for creep suggest that columns with

$$\frac{kL}{r}\sqrt{\frac{N}{f'_c A_g}} > 20 \qquad (1.27)$$

can be considered slender and those for which this ratio exceeds 70 should not be permitted. In this expression, r is calculated for a transformed uncracked section based on a modular ratio of 20.

1.4.6 Shortcomings of Traditional Slender Column Design

As pointed out earlier, the current design procedures were derived for the hypothetical case of elastic hinged columns and then extended using a series of approximations to apply to reinforced concrete columns in frames. Each of these approximations has introduced inaccuracies into the process. The most serious of these are as follows:

(a) No single EI value or $1/\rho$ value is applicable for material failures and stability failures. The CEB Model Column method avoids this problem but is tedious to apply. The use of curvature at the balanced load to define EI or $1/\rho$ shows considerable promise.

STABILITY OF REINFORCED CONCRETE BUILDING FRAMES   23

(b) There is no agreement about how or when to account for creep. The creep eccentricity method shows considerable promise, especially if, as is usually the case, the eccentricities of the sustained and transient loads differ.

(c) The use of effective length factors to account for all the effects of frame action is crude at best. Effective length factors are normally only available for fully braced or fully sway frames. While the first case frequently occurs, fully sway frames are rare. Normally, when analyzing a column in a sway frame, both the gravity load moments and the sway moments are magnified. In this process the gravity load moments are generally amplified too much.

(d) Biaxial bending of slender columns is generally poorly handled.

(e) The definition of when columns are short and slender needs more study. Probably three bounds are needed, one for hinged-end columns, one for restrained columns in braced frames, and a third for columns in sway frames.

(f) Torsional instability of three-dimensional structures is seldom if ever considered.

1.5 DESIGN OF SLENDER REINFORCED CONCRETE COLUMNS — RECENT DEVELOPMENTS

In the past decade several significant new concepts concerning column design have been developed. Four of these will be discussed here. Two apply to sway frames and partially braced frames, the other two apply to braced frames.

1.5.1 Separation of Lateral Drift Effect and Member Stability Effect

The moments and forces in the laterally deflected column in Fig. 1.14(a) can be resolved as shown in Fig. 1.14(b). Columns in frames undergo two types of displacements which lead to slenderness effects. First, the top of the column may displace laterally relative to the bottom of the column. The gravity loads acting on the displaced structure give rise to increased end moments in the columns as illustrated in Fig. 1.8 and eqn (1.2). The increase in storey moment due to the $\Sigma N\Delta$ term will be referred to as the 'lateral drift effect'.

The second type of slenderness effect occurs due to displacements of the column relative to the straight chord line joining the ends of the displaced column, as shown in Fig. 1.14(c). These displacements

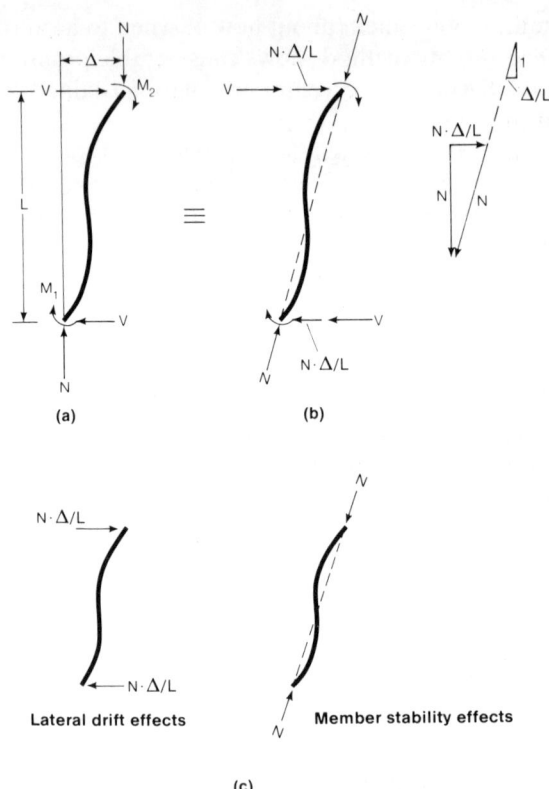

FIG. 1.14. Separation of lateral drift effect and member stability effect. Reproduced from Lai and MacGregor (1983) by permission of the American Society of Civil Engineers.

interact with the axial load in the column to produce further deflections of the column relative to the chord line. The resulting increases in moment will be referred to as the 'member stability effect'.

The lateral drift effect and member stability effect are present to different degrees in different types of frames. In the case of hinged columns, the columns themselves experience only the member stability effect. The lateral drift effect must be entirely resisted by the bracing members which hold the ends of the columns in position. In the case of a sway frame, the columns must resist both the lateral drift and member stability effects. Generally in sway frames the lateral drift effects overshadow the member stability effects. Columns in braced frames fall between. For a frame with very stiff bracing, member stability effects dominate and the lateral drift effects are primarily in the bracing. On the other hand, a frame with very flexible bracing may have significant lateral drift effects combined with member stability effects.

Thus, a 'braced frame' can be defined as a frame in which the columns are not subject to significant lateral drift effects. In such a case only member stability effects are important. In 'unbraced frames' or 'partially braced frames' the columns experience both lateral drift effects and member stability effects.

1.5.1.1 *ACI Code Procedure*
Based on tests of single storey, multi-column frames subjected to beam loads, column loads and lateral loads, Ford *et al.* (1981) proposed that, in the design of unbraced frames, the moments due to loads not causing appreciable sway be magnified using a braced frame moment magnifier (eqns 1.11 and 1.22), and the moments due to loads causing sway be magnified using a sway frame moment magnifier (eqn 1.23). Appreciable sway was taken as Δ greater than $L_u/1500$. The column is then designed for the sum of the two magnified moments. This procedure has been incorporated in the 1983 ACI Code.

1.5.1.2 *Canadian Code Procedure*
Lai and MacGregor (1983) observed that the maximum magnified non-sway moments may occur at a different section in the column than the maximum magnified sway moments. Based on this work the 1984 Canadian Code (Canadian Standards Association, 1984) suggests the following design procedure:

1. Evaluate the moments M_{ns} at each end of the column 'due to loads that cause no appreciable sway'. For most frames, dead and live loads will not cause appreciable sway.

2. Evaluate the lateral load moments M_s at each end of the column, and magnify these for lateral drift effects. The magnified moments $\delta_s M_s$ can be calculated using an exact or approximate second-order analysis, or using a sway magnifier similar to the one in the ACI Code.

3. Add the moments from steps 1 and 2 at each end of the column, designating the numerically larger sum of moments as M_2 and the numerically smaller as M_1.

4. Check whether the moments between the ends of the columns exceed those at the ends. This is done by magnifying the moments from

step 3 to account for member stability effects using the magnifier δ for the braced column case (eqns 1.11 and 1.22) with C_m based on M_1 and M_2 from step 3 and N_c based on the braced frame effective lengths. Normally this magnifier is equal to 1·0 indicating that the maximum moments occur at the ends of the columns.

5. In an unbraced frame it is also necessary to check the ratio of $\Sigma N/\Sigma N_c$ for the case of maximum vertical loads, taking N_c as the sway critical load. If this ratio approaches 1·0 the structure approaches the sway buckling frame instability load.

1.5.1.3 *Colombian Code Procedure*
The 1984 Colombian Code contains a very similar procedure to the Canadian Code (see Asociacion Colombiana de Ingenieria Sismica, 1984). Here, the lateral drift effects are calculated using eqn (1.28) presented later in this chapter. The Colombian Code drafts and Lai's work formed the basis for the 1984 Canadian Code clauses on frame stability.

1.5.2 Calculation of Lateral Drift Effects
Two types of calculation procedures have evolved in the past decade. The first involves the use of 'exact' second-order analyses which solve for the lateral drift effects directly. The second general class of calculation procedures involve modifying the results of first-order analyses to approximate the second-order effects.

1.5.2.1 *'Exact' Second-order Analyses*
In an 'exact' second-order analysis the stiffness matrix includes a displacement parameter. These are generally expressed in terms of stability functions or related functions which, in turn, are functions of the axial loads in the members. This requires an iterative analysis, reevaluating the terms in the stiffness matrix after each cycle to account for changes in axial loads. Examples of this type of analysis are given by Aas-Jakobsen and Grenacher (1974), Ellyin (1983) and El-Zanaty and Murray (1983). With the steady development of more efficient computers these techniques are increasingly more practical.

1.5.2.2 *Approximate Second-order Analyses*

(a) Iterative $N\Delta$ analyses. A number of authors have described an iterative calculation of the lateral drift effects in which the $\Sigma N\Delta$ term in

Fig. 1.8 and eqn (1.2) is estimated by applying equivalent shear forces, $\Sigma N\Delta/L$, at the top and bottom of each storey (Wood et al., 1976; MacGregor and Hage, 1977). In the first iteration the $\Sigma N\Delta/L$ shears are based on the deflection Δ_0 from a first-order (conventional) frame analysis. The deflections from the $\Sigma N\Delta_0/L$ shears give rise to a new set of $\Sigma N\Delta/L$ shears which, in turn, give rise to new deflections. After several iterations a good estimate of the lateral drift effects is obtained.

Figure 1.15(a) shows the moment diagram due to the $\Sigma N\Delta/L$ shears. Figure 1.15(b) shows the actual moment diagram due to the interaction of gravity loads and deflections. From moment-area principles it is obvious that the moment diagram in (b) will produce a larger lateral deflection than that in (a). As a result the $\Sigma N\Delta/L$ calculation underestimates the lateral deflections and hence underestimates the lateral drift effects (Rosenblueth, 1967; Stevens, 1967). This can be corrected by increasing the $\Sigma N\Delta/L$ shears to $\gamma \Sigma N\Delta/L$ where γ approaches 1·0 for very flexible supports and approaches 1·22 for rigid end restraints. Lai and MacGregor (1983) propose an average value of $\gamma = 1\cdot 15$; Stevens (1967) proposed $\gamma = 10/9 = 1\cdot 11$.

The iterative $N\Delta$ method has been shown to break down in a number of academic cases (Ellyin, 1983). This is not a problem for structures meeting practical sway indices Δ/L.

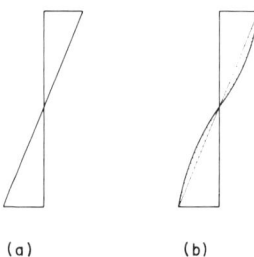

(a)   (b)

FIG. 1.15. Moment diagrams due to $\Sigma N\Delta/L$ shears and $N\Delta$ moments.

(b) Storey magnifier method. In this method the iterative method is simplified by making the additional assumption that each storey behaves independently of other storeys. As a result each storey can be treated like a single-storey frame subjected to the lateral load shears in

that storey, ΣV, plus the modified $N\Delta$ shears, $(\Sigma \gamma N/L)\,\Delta$ of that storey. Because the deflection under horizontal forces only is directly proportional to the applied horizontal force, the following relation is obtained:

$$\frac{\Delta}{\Delta_0} = \frac{\Sigma V + \left(\sum \frac{\gamma N}{L}\right)\Delta}{\Sigma V} \qquad (1.28)$$

in which Δ is the final magnified deflection and Δ_0 is the first-order deflection of the storey. After rearranging the terms, the sway deflection magnifier δ_s, defined as Δ/Δ_0, is equal to

$$\delta_s = \frac{1}{1 - \dfrac{\left(\sum \frac{\gamma N}{L}\right)\Delta_0}{\Sigma V}} \qquad (1.29)$$

in which Σ denotes the summation for all columns in a given storey. Because the moments in a single storey frame subjected to horizontal forces are directly proportional to the deflection, the column end moments are equal to $\delta_s M_{0s}$ where M_{0s} is the first-order end moment in a given column in a sway frame. Expressions similar to eqn (1.29) have been developed by many investigators.

The simplifying assumption of the behavior of a storey being independent of other storeys is reasonable for a frame with flexible columns and stiff beams. Lai and MacGregor (1983) have shown this assumption to be valid subject to two specific conditions: the maximum value of δ_s in the structure should be less than roughly 1·5; and an inflection point should occur at or between the ends of each vertical member, column or wall, in every storey of the structure when it is subjected to lateral loads.

(c) Frame magnifier method. This method is also a simplification of the modified iterative method except that a different simplifying assumption is made. Here it is assumed the deflection ratio Δ/Δ_0 is equal for all the storeys of the frame. In other words, the total lateral deflections of the

STABILITY OF REINFORCED CONCRETE BUILDING FRAMES 29

frame are δ_s times those produced by the lateral loads H, where $\delta_s = \Delta/\Delta_0$. As a result, the energy stored in the structure due to the lateral loads plus the sway forces is identical with the energy resulting from the lateral loads $\delta_s H$. Based on this, the deflection magnifier δ_s becomes equal to

$$\delta_s = \frac{1}{1 - \dfrac{\sum_{i=1}^{n}\left(\sum \dfrac{\gamma N}{L}\right)_i \Delta_{0i}^2}{\sum_{i=1}^{n}(\Sigma V)_i \Delta_{0i}}} \quad (1.30)$$

in which i is the storey level and n is the number of storeys in the structure. Since all the deflections are increased by the same ratio, the column end moments due to lateral loads and lateral drift are equal to $\delta_s M_{0s}$.

The simplifying assumption of equal Δ/Δ_0 in all the storeys has been found to be valid provided the structure includes a distinct shear wall extending from the base to the top of the structure and δ_s is less than roughly 1·5 (Lai and MacGregor, 1983). A 'distinct shear wall' is defined as a stiff vertical element which has fewer than two points of contraflexure.

(d) Accuracy of approximate second-order analyses. Lai and MacGregor (1983) have compared approximate analyses to 'exact' second-order elastic analyses of a series of frames. One of these is summarized in Fig. 1.16. The building is a 24-storey frame with a single concentrated lateral load applied at the top and concentrated vertical loads applied at each joint in the frame and the top of the wall. The stepped lines in Fig. 1.16 give the ratio Δ/Δ_0 from the exact analysis where Δ is the second-order end deflection in the storey and Δ_0 is the corresponding first-order deflection. In all calculations γ was taken as 1·05 except in the bottom storey where 1·20 was used due to the fixed base.

Figure 1.16 shows that the frame magnifier method is a close approximation in the region with the shear wall, and the storey magnifier method applies in the region above the wall.

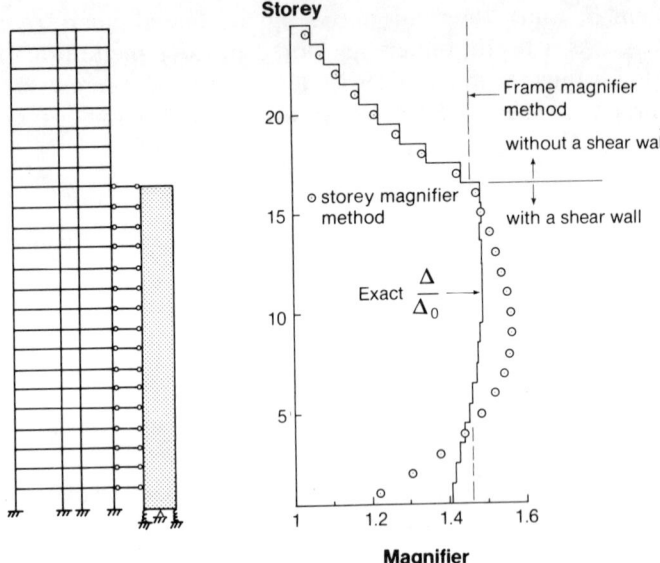

FIG. 1.16. Comparison of approximate and exact solutions — frame with discontinuous wall. Reproduced from Lai and MacGregor (1983) by permission of the American Society of Civil Engineers.

1.5.2.3 *Other Considerations in the Calculation of the Lateral Drift Effects*

(a) Stage of loading. Because there is a non-linear relationship between gravity load and lateral deflection, the analysis must be carried out using loads appropriate to the loading case in question. It is not sufficient to compute second-order moments, etc., using unfactored loads and to then factor and combine these moments.

For serviceability checks the loads and stiffnesses must correspond to those encountered in everyday service. For strength design the loads must be factored and the stiffnesses must correspond to those immediately before failure.

(b) EI values. The stiffnesses, *EI*, must correspond to the loading case under consideration. This may require two or more analyses with different *EI* values and different loads for serviceability and ultimate load situations.

If elastic second-order analyses are to be used in the strength design of a building, it is important that the deflections be representative of those at the factored load level. The EI values for such an analysis should be those at the stage immediately prior to the onset of yielding at the critical sections in the members. Furthermore, the EI values used in computing overall member stiffnesses should reflect the variation of stiffness along the entire length of each member taking into account cracked and uncracked regions and should not merely represent the most highly loaded section. Clearly, however, when dealing with a 20-storey structure with some 1500 members it is seldom economically feasible to do so. A reasonable estimate of EI for second-order analyses at the ultimate limit state can be based on the short time value of E, and $I = 0.4I_g$ for beams and $I = 0.8I_g$ for columns where I_g is the gross moment of inertia ignoring reinforcement (MacGregor and Hage, 1977; ASA, 1984).

The value of EI for shear walls is less well defined since under wind loadings such elements crack only in localized areas of high moment. Frequently the use of $I = 0.8I_g$ will be acceptable although lower values may be indicated in lower storeys.

Lateral loads are not usually sustained loads and hence creep seldom enters the picture in computing lateral drift effects.

At service loads higher moments of inertia (or higher values of the modulus of elasticity E) should be used since the members are less cracked.

(c) Out-of-plumb construction. In real structures, the centroid of the top of a column is often not directly over the centroid at the other end due to construction inaccuracies. As a result, gravity loads acting through the initially inclined column generate additional forces within the structures. Studies of the magnitude of such out-of-plumbs and methods for including them in the lateral drift effects are presented by MacGregor (1979) and Lai and MacGregor (1983).

(d) Resistance factors. When selecting the cross-section of an individual column for the computed loads and moments, resistance factors are used to account for variability in the strength and critical load of the column in question. These factors account for the probability that an individual column may be understrength.

Lateral drift effects, on the other hand, involve not one but many members, beams, columns and walls, the stronger ones carrying some of the moments and lateral shears that would have been assigned to the understrength members. Thus, the overall lateral deflections of a structure are controlled by something approaching the average lateral stiffness of the members. In calculating the lateral drift effects using a second-order analysis or one of the approximate methods given in Section 1.5.2.2, it should be satisfactory to use the specified value of the concrete strength to calculate E if there are more than about eight members, beams and columns participating in resisting the lateral sway. Hence, no resistance factor is required in the calculation of lateral sway effects.

On the other hand, once the lateral drift moments are known, resistance factors should be used in proportioning the individual members.

1.5.2.4 Combination of Second-order Moments and Moments due to Gravity Loads

Depending on the degree of sophistication of the second-order analysis used, it will frequently be necessary to combine the results of the second-order analysis for lateral loads and total vertical loads with the results of a gravity load analysis. This can be done by the method described in Section 1.5.1.2.

1.5.3 Variable Stiffness Method

In Fig. 1.5 it was shown that two things happen simultaneously when a slender column in a braced frame is loaded to failure. First, the reduction in column stiffness due to cracking and axial loads leads to a reduction in the column end moments. Secondly, the column deflects at an increasing rate as the axial load increases. Butler (1977) and Wood and Shaw (1979) have developed the Variable Stiffness Method to consider these two aspects in the design of restrained concrete columns bent in single curvature.

In this method the flexural stiffness of a reinforced concrete column is expressed as $R(EI)_i$ where $(EI)_i$ is the stiffness of an uncracked column based on the initial tangent moduli for the concrete and steel and R is a reduction factor to account for cracking and inelastic action in the concrete. For any given combination of axial load and moment

$$R = \frac{M}{(1/\rho)} / (EI)_i \qquad (1.31)$$

Thus, R is equivalent to the height of the surface in Fig. 1.12 corresponding to the given combination of axial load and moment. Wood and

STABILITY OF REINFORCED CONCRETE BUILDING FRAMES 33

Shaw (1979) developed an approximation for R for calculation purposes.

In the variable stiffness method a reduced EI value R $(EI)_i$ is used in a moment distribution analysis of a frame subassemblage to get the column end moments including the effect of the reduced flexural stiffness of the column on these moments. In the original development, these end moments were then magnified using eqn (1.22) to get the moments at mid-height where N_c was based on the initial EI and the effective length of the column (Butler, 1977). Wood and Shaw (1979) multiplied the end moments by a magnifier obtained graphically. The graphs were derived to account for the effects of stability functions on the end moments in addition to the required moment magnifier.

The significance of the variable stiffness method comes from its recognition that the starting point for the computation of the magnified moments should be the column end moments based on a reduced flexural stiffness. Extremely good accuracy was obtained when compared to Cranston's (1972) computer simulations of column behavior.

1.5.4 Distinction Between Column Moments due to Imposed Loads and those due to Imposed Deformations

In the columns shown in Fig. 1.17(a) and (b) the moments at the base and at mid-height are directly related to the applied loads. These members fail when the axial load and maximum applied moment plus slenderness moment correspond to a failure condition. On the other hand, in the braced frame shown in Fig. 1.18(a) and (b) the column moments all result from the fact that the columns must undergo

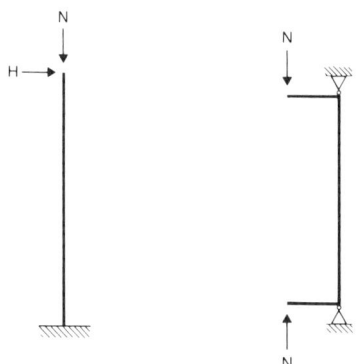

FIG. 1.17. Column moments due to imposed loads.

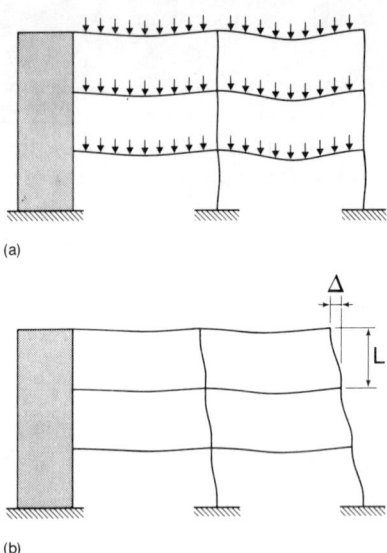

FIG. 1.18. Column moments due to imposed deformations. Reproduced from Favre et al. (1984) by permission of the International Association for Bridge and Structural Engineering.

imposed end rotations to maintain compatibility. In short, the column moments in Fig. 1.17 are required for equilibrium and cannot be dissipated by cracking or inelastic action while those in Fig. 1.18 are not required for equilibrium and hence can be dissipated. To some extent the behavior of columns in Fig. 1.18 is analagous to the column in Fig. 1.5 where the end moment was reduced by inelastic action.

Thürlimann (1984) and Favre et al. (1984) have proposed a design procedure that consists of two stages. First, the imposed deformations at service loads are computed. For the case shown in Fig. 1.18(b) the column is subjected to an imposed end rotation of $\theta = \Delta/L$. These imposed rotations are then compared to limiting service and deflection angles corresponding to excessive crack widths in the plastic hinge area. A subsequent check is carried out at the ultimate limit state to ensure that the imposed end rotation is less than that corresponding to failure of the section. The limiting angle at the ultimate limit state is a function of the ratio of the axial load to the pure axial load capacity, the slenderness of the column and the confinement of the concrete in the hinging regions.

STABILITY OF REINFORCED CONCRETE BUILDING FRAMES 35

This design method is an attempt to simplify column design in braced frames. With proper limits it may be possible to ignore column moments in the design of braced frames, thus reverting to design practice of 50 years ago. It warrants very close study and extension. Related studies of the behavior of columns subjected to imposed deformations have been reported by Grenacher (1976), Hellesland and Scordelis (1981) and Ford *et al.* (1981). Although the Thürlimann–Favre work was limited to columns in braced frames, the dissipation of column moments due to column hinging was noted in tests of sway frames reported by Ford *et al.* (1981) provided there was a column able to resist the moments given up by the softening columns.

1.6 CRITICAL LOADS OF UNBRACED OR PARTIALLY BRACED FRAMES

As the axial load N acting on a hinged column approaches the Euler buckling load N_E, the deflection magnifier δ_D for that column approaches infinity (see eqn 1.4). In the same way, the critical load of a sway frame is that load ΣN for which the denominator of eqn (1.29) is equal to zero. Thus

$$(\Sigma N)_{cr} = \frac{\Sigma VL}{\gamma \Delta_0} \qquad (1.32)$$

where $(\Sigma N)_{cr}$ is the total column load corresponding to the critical load of a storey in the frame, ΣV is the first-order shear in all the columns of the storey, L is the height of the storey, γ is a factor about 1·15, and Δ_0 is the first-order deflection of the storey. For a frame containing a shear wall running the full height of the structure a similar result can be obtained from eqn (1.30).

Stevens (1967) used eqn (1.32) to illustrate a number of very important practical aspects of frame stability. The first of these is that for a given storey shear ΣV the critical load is inversely proportional to the drift angle Δ_0/L. This is of importance in the design of long narrow buildings. For such a building the wind forces on the narrow end are a fraction of those on the wide side. If the same sway index, Δ_0/L, is used in the design for wind forces in the two directions, the critical load for buckling perpendicular to the narrow end will be a fraction of that for buckling in the other direction. As a result, the moment magnification will be larger for moments due to wind forces on the narrow end than for those on the wide side. A major American tall building had to be stiffened in the long direction after completion for this reason.

Based on work by MacGregor and Hage (1977), the Commentary to the ACI Code uses the stability factor Q to define whether frames are braced or unbraced:

$$Q = \frac{\gamma \Sigma N \Delta_0}{\Sigma V L} \qquad (1.33)$$

where γ is taken to be 1·0. For a frame that is free to sway, Q is a close approximation to $\Sigma N / \Sigma N_{cr}$. Equation (1.29) suggests that if Q is less than 0·04 for a storey, the total magnified moments will not exceed 105% of the first-order moments and the structure can be considered braced.

The 1984 Colombian Code uses Q in a similar manner. Storeys with a stability index Q less than 0·1 are considered braced. Those with Q between 0·1 and 0·3 can be designed using the process described in Sections 1.5.1.2 and 1.5.1.3 of this chapter. Structures with Q greater than 0·3 require an exact second-order analysis and those with Q greater than 0·5 are considered 'unstable' for practical considerations and should be stiffened or braced.

Horne (1975) has presented a method for estimating the critical buckling load factor λ_c for a frame, where λ_c is the factor by which the gravity loads at each floor W_i must be multiplied to cause buckling. A first-order analysis is carried out for fictitious lateral loads of $0·1 W_i$ at each floor level. The critical load factor is then calculated from the largest calculated ratio of Δ/L in all the storeys in the building:

$$\lambda_c = \frac{0·009}{(\Delta/L)_{max}} \qquad (1.34)$$

Horne has shown that this underestimates the critical load by 0–20% and for beam and column frames is surprisingly accurate. Horne's checks did not include buildings with shear walls. Equation (1.34) is merely eqn (1.33) restated with $\lambda_c = 1/Q$.

The 1982 draft of the British Steel Structures Code classifies frames as either 'non-sway' or 'sway' frames based on their lateral deflections when subjected to fictitious lateral loads of $0·005 W_i$ at each floor level. The close relationship between this procedure, the ACI Commentary procedure based on eqns (1.32) and (1.33) and Horne's method can easily be seen.

1.7 TORSIONAL STABILITY OF THREE-DIMENSIONAL BUILDINGS

Although all buildings are three-dimensional, the possibility of torsional instability is seldom considered in design. Figure 1.19 represents a three-dimensional building. As the top of a storey in this building undergoes a rotation ϕ about the center of resistance relative to the bottom of the storey, the top of the jth column in any given storey moves ϕr_j horizontally from point 1 to point 2. The axial force N_{ij} in this column now causes an $N\Delta$ moment which can be replaced by the horizontal force $P_{ij}\phi r_j/L$. This force and the similar forces in all the other columns cause a second-order torque about the center of rotation. Wynhoven and Adams (1972) have used this technique in the analysis of three-dimensional steel frames.

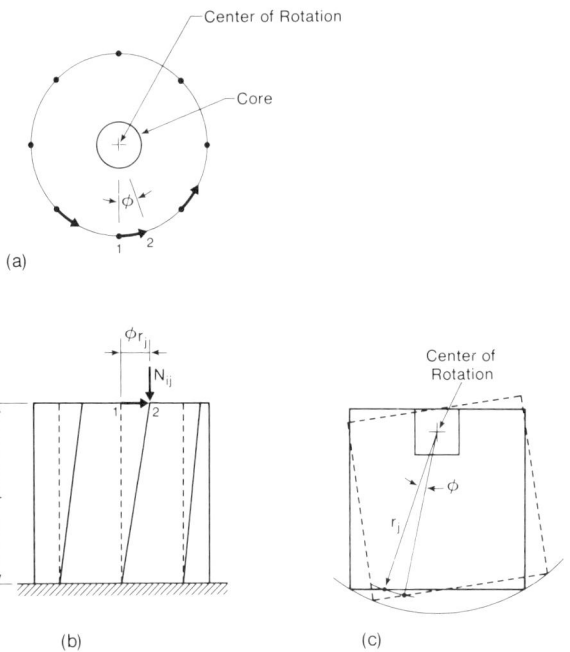

FIG. 1.19. Torsional $N\Delta$ effects in a three-dimensional building frame: (a) plan; (b) elevation; (c) general case. Reproduced by permission of the American Society of Civil Engineers.

Following the method used to derive eqn (1.29), the total magnified torque T can be expressed as

$$T = \frac{T_1}{1 - \dfrac{\Sigma \gamma (N_j r_j^2) \phi_0}{T_1 L}} \tag{1.35}$$

when T_1 is the first-order torque on the storey and ϕ_0 is the first-order rotation caused by that torque.

The magnified torque in eqn (1.35) tends to infinity when the denominator of eqn (1.35) tends to zero. Thus the critical load for torsional buckling of a storey can be stated as

$$\Sigma (Nr^2)_{cr} = \frac{T_1 L}{\phi_0} \tag{1.36}$$

This expression shows two very important things about the torsional stability of a storey. First, the higher the torsional stiffness T/ϕ, the higher the critical load. Secondly, the critical load is affected both by the magnitude of the loads N, and by the square of their distance from the centre of rotation. This suggests that long buildings or cruciform buildings with one central shear core may be particularly susceptible to torsional instability.

Analysis of the torsional buckling load of buildings is discussed by Goldberg (1973). Methods of carrying out three-dimensional second-order analyses are presented by Wynhoven and Adams (1972), Argent and Banut (1979), and Rutenberg (1982).

1.8 CONCLUDING REMARKS

The current design procedures for concrete columns are based on an elastic analysis applicable to single-curvature, hinged-end columns. Approximation after approximation has been introduced to extend this method of analysis to concrete columns and to columns in frames. The most questionable approximations involve the choice of the EI for the concrete section, the treatment of creep, and the use of elastic effective length factors to approximate frame behavior.

Four new design concepts for columns in frames have been reviewed in Section 1.5. All four of these bear further intensive study with a view to extracting the valuable concepts from each to incorporate into a unified design process.

REFERENCES

AAS-JAKOBSEN, K. and GRENACHER, M. (1974) Analysis of slender reinforced concrete frames. *Publications, International Association for Bridge and Structural Engineering*, **34-1**, 1–18.

AMERICAN CONCRETE INSTITUTE (1983) *Building Code Requirements for Reinforced Concrete*, ACI 318-83, 111 pp.

ARGENT, R. and BANUT, V. (1979) *Structural Analysis of Reinforced Concrete with Slender Columns* (in Rumanian), Editura Technica, Bucharest, 264 pp.

ASOCIACION COLOMBIANA DE INGENIERIA SISMICA (1984) *Colombian Code for Seismic Resistant Construction* (in Spanish), 306 pp.

AUSTRALIAN STANDARDS ASSOCIATION (1984) *Draft Australian Standard, Unified Concrete Structures Code*, BD12/84-1, February, 158 pp.

BRITISH STANDARDS INSTITUTION (1972) *Structural Use of Concrete*, CP110, Part 1, 154 pp.

BUTLER, D. J. (1977) Strength of restrained reinforced concrete columns — a new approach. *Magazine of Concrete Research (London)*, **29** (100), 113–22.

CANADIAN STANDARDS ASSOCIATION (1984) *Design of Concrete Structures for Buildings*, CAN3-A23.3-M84, 280 pp.

COMITÉ EURO-INTERNATIONAL DU BÉTON (1978a) CEB–FIP Model Code for concrete structures. *Bulletin d'Information No. 124/125E*, 348 pp.

COMITÉ EURO-INTERNATIONAL DU BÉTON (1978b) *CEB–FIP Manual of Buckling and Instability*, The Construction Press, Lancaster, England, 135 pp.

CRANSTON, W. B. (1972) Analysis and design of reinforced concrete columns. *Research Report No. 20*, Cement and Concrete Association, London.

ELLYIN, F. (1983) Nonlinear response and stability of structures. *Canadian Journal of Civil Engineering*, **10** (1), 27–35.

EL-ZANATY, M. H. and MURRAY, D. W. (1983) Nonlinear finite element analysis of steel frames. *Journal of Structural Engineering, American Society of Civil Engineers*, **109** (2), 353–68.

FAVRE, R., NAJDANOVIC, D., SUTER, R. and THÜRLIMANN, C. (1984) New design concept for reinforced concrete columns in buildings. *Final Report, 12th Congress of International Association for Bridge and Structural Engineers, Vancouver*, pp. 879–86.

FEDERATION INTERNATIONALE DU PRECONTRAINT (1982) *FIP Recommendations on Practical Design*, 76 pp.

FORD, J. S., CHANG, D. C. and BREEN, J. E. (1981) Design indications from tests of unbraced multipanel concrete frames. *Concrete International: Design and Construction*, **3** (3), 37–47.

FURLONG, R. W. and FERGUSON, P. M. (1966) Tests of frames with columns in single curvature. *Symposium on Reinforced Concrete Columns*, ACI Publication SP-13, American Concrete Institute, Detroit, pp. 55–73.

GOLDBERG, J. E. (1973) Approximate methods for stability and frequency analysis of tall buildings. *Proceedings of Regional Conference on Tall Buildings, Madrid*, September 17–19, 1973, pp. 123–46.

GREEN, R. and BREEN, J. E. (1984) Eccentrically loaded concrete columns; 15 years of sustained load. *Final Report, 12th Congress of International Association for Bridge and Structural Engineers, Vancouver*, pp. 911–18.

GRENACHER, M. (1976) *Einfluss von Verschiebungen und verschiedenen Lagerungen auf des Traguerhalten von Stahlbetonstutzen,* Bericht No. 61, Institut für Baustatik und Konstruktion, ETH, Zurich.

GVOZDEV, A. A. and CHISTIAKOV, E. A. (1972) Effect of creep on load capacity of slender compressed elements. *Proceedings of the International Conference on Planning and Design of Tall Buildings,* Vol. III, American Society of Civil Engineers, pp. 537–54.

HELLESLAND, J. (1976) Approximate second-order analysis of unbraced frames. *Internal Report,* Dr. A. Aas-Jakobsen, A/S, Oslo, Norway, 50 pp.

HELLESLAND, J. and SCORDELIS, A. C. (1981) Analysis of R.C. bridge columns under imposed deformations. *Advanced Mechanics of Reinforced Concrete,* IABSE Reports No. 34.

HORNE, M. R. (1975) An approximate method for calculating the elastic critical loads of multi-storey plane frames. *Structural Engineer,* **53** (6), 242–8.

LAI, S-M. A. and MACGREGOR, J. G. (1983) Geometric non-linearities in multi-storey frames. *Journal of Structural Engineering, American Society of Civil Engineers,* **109** (ST11), 2528–45.

LAI, S-M. A., MACGREGOR, J. G. and HELLESLAND, J. (1983) Geometric non-linearities in non-sway frames. *Journal of Structural Engineering, American Society of Civil Engineers,* **109** (ST12), 2770–85.

MACGREGOR, J. G. (1979) Out-of-plumb columns in concrete structures, *Concrete International — Design and Construction,* **1** (6), 26–31.

MACGREGOR, J. G. and HAGE, S. E. (1977) Stability analysis and design of concrete frames. *Journal of the Structural Division, American Society of Civil Engineers,* **103** (ST10), 1953–70.

MACGREGOR, J. G., BREEN, J. E. and PFRANG, E. O. (1970) Design of slender concrete columns. *American Concrete Institute Journal,* **67** (1), 6–28.

MACGREGOR, J. G., OELHAFEN, U. H. and HAGE, S. E. (1975) A re-examination of the EI value for slender columns. *Reinforced Concrete Columns,* American Concrete Institute, Detroit, pp. 1–40.

MANUEL, R. F. and MACGREGOR, J. G. (1967) Analysis of restrained reinforced concrete columns under sustained load, *American Concrete Institute Journal,* **64** (1), 12–23.

MENEGOTTO, M. (1983) Observations on slenderness bounds for R.C. columns. *Bulletin d'Information No. 155,* Comité Euro-International du Béton, Paris, pp. 5–30.

MENN, C. (1974) Simple method for determining the ultimate load of slender compression members (in German). *Preliminary Report, Symposium on the Design and Safety of Reinforced Concrete Compression Members,* International Association for Bridge and Structural Engineering, Vol. 16, pp. 136–44.

ROSENBLUETH, E. (1967) Slenderness effects in buildings. *Journal of the Structural Division, American Society of Civil Engineers,* **91** (ST1), 229–52.

RUTENBERG, A. (1982) Simplified P-delta analyses for asymmetric structures. *Journal of the Structural Division, American Society of Civil Engineers,* **108** (ST9), 1995–2013.

SMITH, R. G. and BRIDGE, R. Q. (1984) Design of concrete columns. *Top-Tier Design Methods in the Draft Concrete Code,* School of Civil and Mining Engineering, University of Sydney, Australia, pp. 2.1–2.95.

STEVENS, L. K. (1967) Elastic stability of practical multistory frames, *Proceedings, Institution of Civil Engineers,* **36**, 99-117.

SWISS ENGINEERS AND ARCHITECTS ASSOCIATION, SIA (1976) Ultimate Load Design for Compression Members, Revisions to Articles 3.08, 3.09, 3.24 of the SIA Code 162 (1968) (in German), SIA, Zurich, 7 pp.

THÜRLIMANN, C. B. (1984) Design of reinforced concrete columns subjected to imposed end deformations. IABSE Proceedings P-77/84, *IABSE Periodica 4/1984*, International Association for Bridge and Structural Engineering, Zurich, pp. 89-108.

WARNER, R. F. and KORDINA, K. (1975) Influence of creep on the deflections of slender reinforced concrete columns. *Deutscher Ausschuss für Stahlbeton,* **250**, 156 pp.

WARNER, R. F., RANGAN, B. V. and HALL, A. S. (1976) *Reinforced Concrete,* Pitman, Carlton, Australia, 475 pp.

WOOD, R. H. and SHAW, M. R. (1979) Developments in the variable-stiffness approach to reinforced concrete design. *Magazine of Concrete Research (London),* **31** (108), 127-41.

WOOD, B. R., BEAULIEU, D. and ADAMS, P. F. (1976) Column design by P delta method, *Journal of the Structural Division, American Society of Civil Engineers,* **102** (ST2), 411-27.

WYNHOVEN, J. H. and ADAMS, P. F. (1972) Analysis of three-dimensional structures. *Journal of the Structural Division, American Society of Civil Engineers,* **98** (ST7), 1361-76.

Chapter 2

STABILITY OF COMPRESSION MEMBERS

U. QUAST
Arbeitsbereich Massivbau, Technische Universität, Hamburg-Harburg, Federal Republic of Germany

SUMMARY

The particular stability problem of reinforced concrete members which arises from stiffness reduction due to increasing bending is described. The difference between this and buckling with equilibrium bifurcation is explained. The principles for checking stability as formulated in the Eurocode 2 (1984) are discussed, as well as the fundamental ideas for a direct design method. Some simplifications for special members are mentioned, viz. chimneys, sway frames, cantilever columns, columns in braced frames. An example is given for a simplified design procedure. Brief reference is made to the so-far incompletely solved problems of biaxial bending and stability of slender walls.

NOTATION

A	Area of cross-section
a	Deflection
$a.\mathrm{I}$	First order deflection
$a.\mathrm{II}$	Second order deflection
b	Width of cross-section
C	Compression force in the compression zone
c	Cover of the reinforcement
E	Modulus of elasticity, Young's modulus
EI	Bending stiffness
e	Eccentricity
$e.\mathrm{I}$	Eccentricity arising from first order theory (no deflection effects)

$e.\text{II}$	Eccentricity arising from second order theory
F	Force, load
f	Strength
H	Horizontal force
h	Total height or diameter of cross-section
i	Radius of gyration
l	Member length
M	Bending moment
$M.\text{I}$	Bending moment arising from first order theory (no deflection effects)
$M.\text{II}$	Bending moment arising from second order theory
m	Relative bending moment
N	Axial force
n	Relative axial force or exponent
r	Radius of the curved member (curvature $= 1/r$)
s	Sign coefficient
T	Tensile force of the reinforcement
V	Vertical force
x, y, z	Coordinates
α	Angle, tilt, slope, inclination
γ	Safety factor
ε	Strain
η	Coefficient for modifying the stress–strain diagram for concrete in order to take creep into account
λ	Slenderness ratio, $\lambda = l:i$
ρ	Curvature coefficient
ψ	Unintentional or accidental slope or tilt
ω	Mechanical ratio of reinforcement

Subscripts

a	Member end (a) or accidental, unintentional
b	Member end (b)
c	Concrete
d	Design
E	Euler
F	Force
k	Characteristic
m	Material
p	Prestress

STABILITY OF COMPRESSION MEMBERS

s   Steel
t   Tension
u   Ultimate
y   Yielding

2.1 INTRODUCTION

The stability problem of slender reinforced concrete members has only been fully studied and understood through the use of computers. For this reason the early regulations prior to 1960 are inadequate when assessed from today's standpoint. The early design procedures took the buckling effect into account by designing or checking the cross-sections for an increased axial load. The next stage was to use magnifiers on the moments alone, which approximate better with real member behaviour. A perfect solution is possible today using computers which need be only the so-called micros. It is expected that the direct use of computers will become common practice, not only for large members such as bridge-piers, towers and chimneys, but also for prefabricated units in standard structures. The problems which now remain are of a fundamental nature: what level of safety factors should be used; what stress–strain curves are applicable; at what slenderness ratio should long column effects be considered; what simplifications are reasonable for calculations done by hand and what checks should be applied to computer results; and so on.

One of the simplest types of compression member involved with slenderness effects is the cantilever column clamped at its base. It is therefore chosen in this chapter for initial discussion. (For the same reason it is defined as the standard member in the CEB/FIP (1978b) Model Code and called 'model column'.) The practical task in reinforced concrete engineering consists of determining the necessary amount of reinforcement for a chosen cross-section rather than checking the safety for a cross-section. For the deflected section of the column in Fig. 2.1, this means determining the tension force T from the moment equilibrium of the other forces V and H related to the position of the compression stress resultant C. The strain at the tension fibre has to be limited to the yielding strain, otherwise the column deflection, and with it the acting moment, increases without increasing the stabilising tension force T. Therefore in many cases a stable equilibrium is not possible for tensile strains greater than the strain at yielding. For an

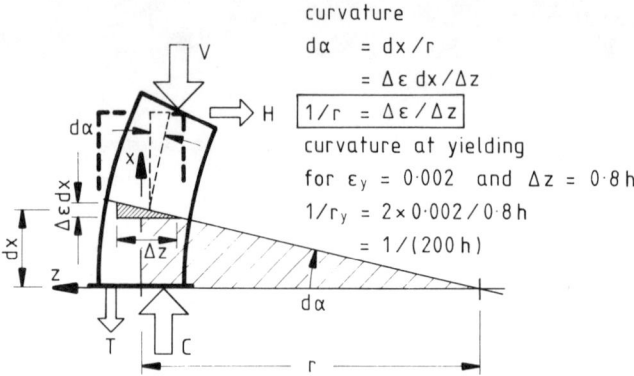

FIG. 2.1. Strains, curvature and displacement.

economical design the maximum tensile force T should correspond to the onset of yielding, and consequently the curvature $1/r$ can be derived directly from the strains without calculating beforehand the effective bending stiffness EI for the cracked cross-section and then finding $1/r$ by dividing the bending moment M by EI. This fundamental idea leads to the direct design method, which is more suitable for concrete members than other methods developed for steel structures. The comparison given in Table 2.1 shows that fewer steps are required in the process and, in particular, iteration is not necessary.

From Fig. 2.1 it will also be seen that the vertical force V only acts unfavourably regarding the necessary reinforcement if the deflection is large. On bridge-piers with box sections the resultant vertical force including dead load of the pier often acts favourably. This departure from classical buckling theory arises as a consequence of the small tensile strength of concrete. This phenomenon may lead to the idea of prestressing such types of compression members, which, although possible, has not yet been adopted in practice apart from special cases. When considering the compression force C and the required concrete area, vertical loads always act unfavourably.

Other reasons why the stability problem of reinforced concrete members needs special treatment are listed in Table 2.2.

The resistance of the column can be expressed as the resistance of its most stressed section ((a) in Fig. 2.2) in terms of the resisting moment, res.M, which is the stress resultant calculated from the cross-section under the curvature $1/r$. The stress resultant N has always to equal the

STABILITY OF COMPRESSION MEMBERS

TABLE 2.1
COMPARISON OF DESIGN METHODS FOR SLENDER REINFORCED CONCRETE MEMBERS

Stiffness method	Direct design method
(1) Assume the amount of reinforcement	
(2) Assume an effective stiffness	Greatest curvature arises at yielding of the reinforcement either at the tension or compression side
(3) Deflection follows from second order theory	Deflection follows from greatest curvature
(4) Take account of second order moment	Take account of second order moment
(5) Compute necessary reinforcement	Compute necessary reinforcement
(6) Repetition from step (1) if the amount of steel obtained in step (5) does not agree with that assumed at step (1)	

TABLE 2.2
IMPORTANT DIFFERENCES BETWEEN STEEL STRUCTURES AND REINFORCED CONCRETE STRUCTURES WITH REGARD TO STABILITY

Steel structures	Reinforced concrete structures
(1) Ultimate moment decreases with increasing axial force, axial force is unfavourable	Ultimate moment may increase with increasing axial force, axial force may be favourable
(2) Bending stiffness is constant	Bending stiffness decreases due to cracking and depends on the ratio of reinforcement

acting axial force V, which is constant for practical purposes and can always be attained by a certain axial strain, the value of which is not required for this discussion. The relationship between the resisting moment, res.M, and the curvature, $1/r$, gives a typical nonlinear relation. The response of the column in terms of the acting bending moment, act.M, including second order effects, comes out as a straight line if the simplifications in Fig. 2.2 are applied. Equilibrium means intersection

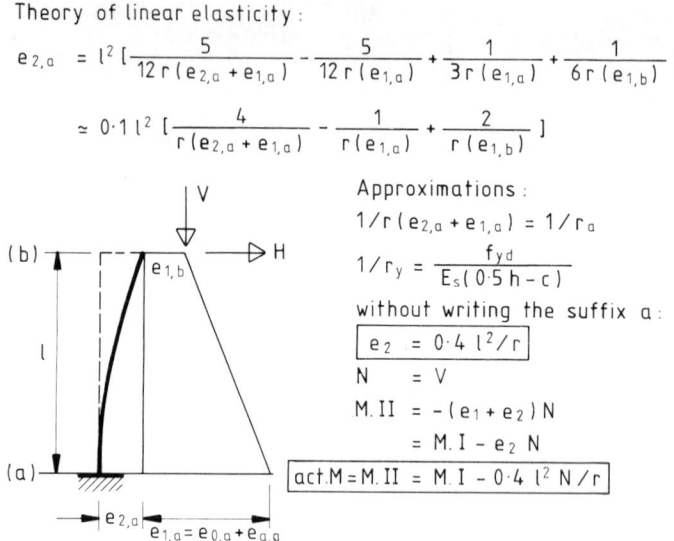

FIG. 2.2. Model column.

or at least contact in one point (e) of these two $M - 1/r$ lines. Following the CEB design philosophy, explained later, the admissible moment, adm.M, for any local cross-section is calculated from the characteristic value of the concrete strength, which is smaller than the average value and which is taken for determining the global deformation behaviour of the member. This is why the res.M line in Fig. 2.3 is curtailed. The second order effect is included in the act.M line, showing the moment magnification with increasing curvature. The inclination of this line depends on the square of the column length l. Depending on the length l and the first order moment M.I, the point (e) of equilibrium may be the end of the res.M line, corresponding to attaining the capacity of the cross-section. With increasing length and decreasing first order moment the point (e) is anywhere at the res.M line, the act.M line being tangent to it. In such a situation the ultimate state of the member is not governed by the cross-section capacity but by the loss of stability due to stiffness reduction with increasing curvature. Stiffness may be defined as the gradient $dM/d(1/r)$. This particular stability problem arises only with very slender members and even in such cases, will not arise if a large accidental eccentricity is to be taken into account in design.

STABILITY OF COMPRESSION MEMBERS

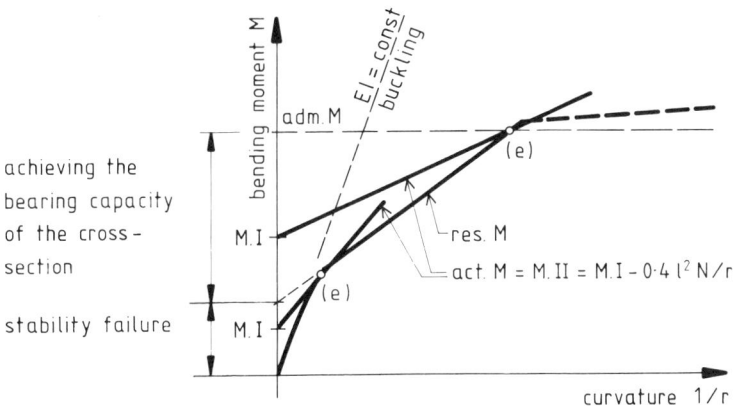

FIG. 2.3. Limit states of equilibrium for the model column.

This general way of considering the stability problem can also deal with the problem of buckling under a centrally applied load. With a zero first order moment the act.M line is tangent of the res.M line in the origin with the gradient $dM/d(1/r) = M/(1/r) = EI$. The simplified equation from Fig. 2.3 becomes

$$M.\text{II} = M.\text{I} - 0.4 * N * l^2/r \qquad (2.1)$$
$$M.\text{II} = 0 \quad - 0.4 * N * l^2 * M.\text{II}/EI$$

and has two solutions: $M.\text{II} = 0$ for any value of N and crit.$N = -10EI/(4l^2)$ for any value of $M.\text{II}$, the latter being the expression for the well known Euler or buckling load, with the approximation $\pi^2 = 10$. For crit.N the resulting moment $M.\text{II}$ is unlimited, which means that the member collapses and at least theoretically this collapse occurs suddenly by equilibrium bifurcation when the loading reaches crit.N. In order to distinguish the above mentioned stability failure due to loss of stiffness from the buckling failure with equilibrium bifurcation, it is also named 'stability failure without equilibrium bifurcation'.

2.2 PRINCIPLES FOR THE STABILITY CHECK

The principles for the stability check in the Eurocode 2 (1984) are adopted from the CEB/FIP (1978b) Model Code, which has received wide international recognition. These principles have already been adopted as a basis in the present regulations of many countries, but certain differences do arise.

2.2.1 Safety Factors

The semi-probabilistic approach using partial safety factors for calculating the design effects of the actions and the design resistance of the structure or its cross-sections in the ultimate limit state from their characteristic values is a good compromise between a high level reliability theory and a simple method as it is demanded for practical use. In contrast to other structural elements, the splitting of the global safety factor into partial factors, which may be different for different actions and different materials, is of great influence on the design of compression members. For this reason the load factor should be further divided into a factor for the loads and a factor for the uncertainty of the calculation model. This avoids a conservative design. If doing so, it is not always sufficient to multiply the calculated effects of the factored actions by the factor for the model uncertainty and compare them with the capacity of the cross-section corresponding to factored material strengths. If, however, the problem is governed by stability failure due to stiffness reduction, without achieving the ultimate capacity of the sections, it is also necessary to check the equilibrium for action effects due to increased deflections by multiplying them by the safety factor for the model uncertainty. Thus, for simplicity and apart from exceptional structures, the splitting of the load factor is not adopted. When drafting the new German standard for reinforced or prestressed concrete poles, the great influence of using partial safety factors was confirmed, as was reported by Quast (1984). Only on the basis of this method was it possible to show adequate safety for all types of poles and antenna masts as they had been successfully constructed in the past; it was impossible to do so with the regulations in the present German reinforced concrete standard, which still uses the concept with a single global load factor. Table 2.3 gives the values for calculating the action effects.

As already explained with Fig. 2.1 the vertical load may act favourably or unfavourably, depending on its point of application in relation to the compression stress resultant C in the concerned section. This may lead to the situation that the dead load of a certain part of the structure may act favourably in one section and unfavourably in another section at the same time. What factor shall then be taken? In order to

STABILITY OF COMPRESSION MEMBERS 51

TABLE 2.3
SAFETY FACTORS FOR ACTIONS IN THE ULTIMATE LIMIT STATE

Actions	Unfavourable effect	Favourable effect
(1) Permanent	1·35	1·0
(2) Prestress	1·2	0·9
(3) Variable	1·5	0·0

avoid unnecessary complications it should be said in the regulations that structures as a whole are only to be analysed first assuming all the permanent load acts unfavourably and secondly that all the permanent load acts favourably.

The action effects are determined for so-called representative actions. This means that several variable actions which are independent from each other are combined in such a way that only one action is taken with its maximum or characteristic value and all the remaining actions are reduced by a so-called combination factor. In the CEB/FIP (1978b) Model Code the combination factors range from 0·6 for parking areas, offices and retail stores to 0·3 for dwellings, whereas Eurocode 2 (1984) prefers a constant combination factor of 0·7 for all types of variable actions. In general, various combinations are to be considered in order to find out the most unfavourable. This can lead to lengthy calculations, which are only justified for exceptional structures. For most structures a simplified approach is sufficient. Eurocode 2 (1984) therefore proposes to use a constant mean safety factor of 1·43 for permanent and variable actions.

2.2.2 Stress–Strain Relationships

The design resistance of structures or cross-sections is based on design values for the material strengths, which are obtained by dividing the characteristic values by the appropriate safety factor. It is normally 1·5 for concrete and 1·15 for steel. For the stability check of a structure as a whole, however, it is permissible to calculate the deflections with a factor of only 1·2 for concrete. The reason is that the overall deflection results from the mean concrete strength and not from the most unfavourable value, being present only in some sections and not in the whole structure. Dividing the characteristic value by 1·2 yields approximately the same design value as dividing the mean value by 1·5. The stress–strain diagrams are shown in Fig. 2.4.

For determining the ultimate capacity of cross-sections one uses the parabolic-rectangular diagram for concrete whereas the deflections are calculated with a curve, which gives a more accurate value for the initial modulus of elasticity and does not include the strength reduction of 0·85 with respect to sustained loading. This corresponds better with real member behaviour but it prevents the establishment of dimensionless design charts. In contrast to the concrete diagrams the safety factor for steel is applied only to reduce the strength.

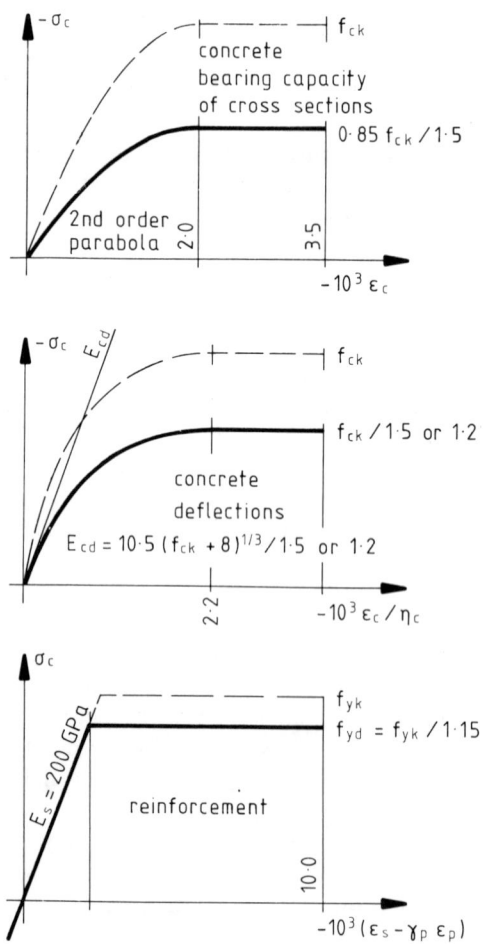

FIG. 2.4. Stress–strain relationships.

The concrete modulus of elasticity is not so closely correlated to the strength as the given formula in Fig. 2.4 may suggest. However, to take care of possible differences due to different types of aggregates, it is prudent to reduce the modulus of elasticity by the amount suggested.

Creep effects can be taken into account either by augmenting the load eccentricity by an additional creep-eccentricity or by multiplying the strains with an appropriate coefficient η_c. The latter method is more suitable for calculations on the basis of stress–strain relationships using computers.

2.2.3 Tension Stiffening Effects

Eurocode 2 (1984) does not permit the use of the tension stiffening effect of concrete, although this can be very important as can be seen from Fig. 2.5, which is based on experiments by Schlaich *et al.* (1979). The different parts of the measured moment–curvature relationship belong to different directions of bending. They had been rotated up to 180° by the same sample. As can be seen, the original line is continued after reloading. Thus one can state that one always obtains the same curvature when reaching a moment for the first time and that it is not affected by load reversals which might have occurred before on a lower load level. This is very important because the ultimate limit state can, by definition, only be reached once and therefore it cannot be affected by load reversals which occur under service load conditions. Therefore it is appropriate to take tension stiffening effects into account for calculating

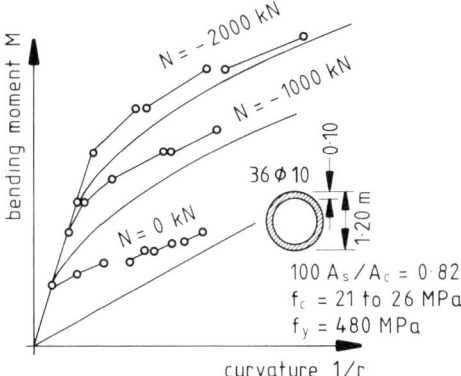

FIG. 2.5. Measured curvatures compared with calculated (not taking tension stiffening effects of the concrete into account).

the deflections in the ultimate limit state and this should be allowed in regulations.

Rao (1966) has presented early work on this effect. Rabich (1969) reported on a simplified method based on linearisation of the moment-curvature interaction between the points of cracking and yielding. This procedure works well for rectangular cross-sections, especially with concentrated reinforcements near the edges. It can be seen from Fig. 2.5 that it is less accurate for circular or tubular cross-sections. More recently, many approaches have been published, especially in connection with computer methods. Some approaches for practical use are mentioned below.

Gilbert and Warner (1978) have suggested increasing the steel area to an effective value resulting directly in reduced tensile strains. Noakowski and Kupfer (1981) propose calculating deflections with a reduced steel strain depending on bond characteristics and the strain value in the crack. Quast (1977) used tension stiffening effects for calculating the member deflection by defining mean values for the concrete tensile stresses between the cracks. This method is described by Quast (1981) and is illustrated in Fig. 2.6. The basic idea is to enlarge the use of well-known methods from engineering practice. Therefore the same type of curve is used for compression as well as for tension stresses. It is a parabolic-rectangular diagram but with a parabola with exponent

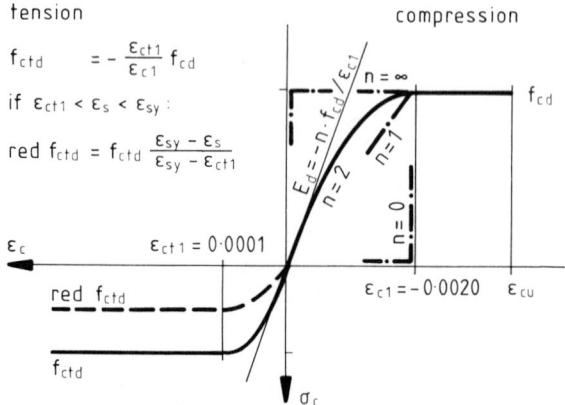

FIG. 2.6. Stress–strain relationship of concrete for determining the mean curvature of reinforced concrete structures (taking tension stiffening into account).

STABILITY OF COMPRESSION MEMBERS 55

n. The exponent n can be chosen to give any design value for the initial modulus of elasticity as can be seen from the given formula. As shown in the figure, for some integer values of n any stress–strain relationship may be defined by this approach. The necessary expressions for computing the stress resultants are only slightly more complex than for the standard diagram with the exponent $n = 2$. Tension stiffening vanishes when the reinforcement yields. This is simply included in the given formulae. The given constant strain values were found by comparisons with test results comprising beams with different shapes of cross-sections and also including prestressing. Creep can be included by increasing the strains, as already mentioned.

The formulation in Fig. 2.6 is not an attempt to define the real concrete behaviour which is, in fact, extremely complex. Nevertheless it gives a good agreement with real deformation behaviour of reinforced concrete structures. It has the advantage that it is only a small modification to established procedures.

2.2.4 Geometrical and Structural Imperfections

To allow for geometrical inaccuracies and to provide additional safety, an additional eccentricity has to be introduced in the most unfavourable direction. The CEB/FIP (1978*b*) Model Code suggests 1/300 of the effective member length but not less than 20 mm. For sway frames the additional eccentricity can be replaced by an unintended slope from the vertical of $\psi = 1/150$ for single storey frames and structures loaded mainly at the top. For other frames a slope of $\psi = 1/200$ is sufficient. These regulations from the CEB/FIP (1978*b*) Model Code are adopted in the Eurocode 2 (1984). It is emphasised, that the additional eccentricity is not just to take account of construction tolerance, but to provide an additional safety reserve.

In the present discussion of the drafts for the Eurocodes for different materials, an attempt is being made to write common model chapters for some main principles, for example imperfections, which have to be taken into account for stability checks. In a framed structure as shown in Fig. 2.7 different geometrical imperfections may occur which can be measured. Their statistical distribution can be determined and a characteristic value can be fixed for regulating purposes. Shear walls or lift shafts may deviate from the vertical in each storey. Arising from checks during construction, continual corrections are taken, giving a final overall slope from the vertical, which is to be expected to decrease with increasing height. Individual columns may also be curved, crooked

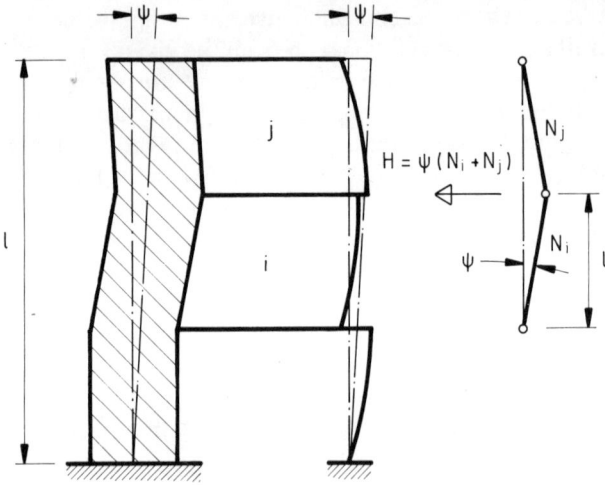

FIG. 2.7. Representation of geometrical imperfections by an unintentional slope ψ.

and their ends may be displaced from one storey to the other. These inaccuracies cause additional actions in the columns which are more important with increasing slenderness. For the design of the structure as a whole, an unintended slope of all the columns should be taken into account. For the connection of the slabs to the shear walls, additional forces H have to be considered, resulting from inclinations opposite to each other in adjoining storeys. The total force for a greater number of columns may be reduced because the mean value of out-of-vertical statistically decreases with an increasing number of elements. This action normally needs no special consideration with structures concreted *in situ* but may be of importance with precast structures. Finally the stability check of a single column can be carried out in terms of a bent bar or an axial load with an additional eccentricity. (These comments are valid for vertical members as well as for those in any direction.)

The effects of inclined members can be taken into account by adding horizontal forces to each vertical force. This procedure can be generalised to a universal procedure for considering imperfections. A certain part of each axial load has to be added as a transversal load in the most unfavourable direction. This is shown in Fig. 2.8. The

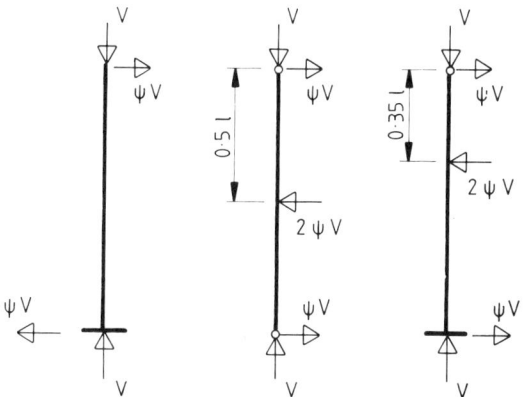

FIG. 2.8. Lateral forces as a procedure for considering actions of imperfections.

equilibrium condition for columns with two fixed ends requires an additional force between the ends which then produces the additional bending action. The unanswered questions are: From which member length shall the unintended slope begin to be reduced and by which expression? How shall the additional horizontal forces be reduced for the combination of a greater number of columns? An existing proposal for member lengths $l > 5\cdot 0$ m and n columns is

$$\psi = \sqrt{5\cdot 0/l} \times \sqrt{(1 + 1/n)/2}/250 \qquad (2.2)$$

In addition to geometrical imperfections, structural imperfections have to be considered. For steel structures these are mainly material inhomogeneities and eigen-stresses. A practical solution is to increase the geometrical imperfection as an alternative. A proposal for steel structures is to substitute the denominator 250 in eqn (2.2) by 200. For concrete structures there are no definitive ideas as to how to treat structural imperfections. A deviation between the mechanical and the geometrical axis, creep under permanent load (if neglected in the calculation for simplification) and structural defects may cause changes in the moment–curvature relationship as shown in Fig. 2.9. The differences between the idealised line and the dotted line with imperfections can be substituted by an additional eccentricity which depends on the slenderness. The background of this interpretation was given with Fig. 2.3. This problem has not been studied intensively and therefore it is not clear if an additional eccentricity substituting for

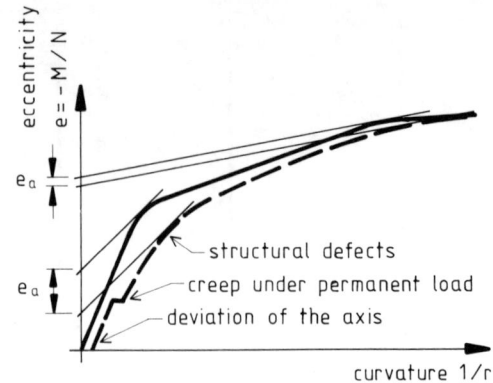

FIG. 2.9. Effects of different types of imperfections.

structural imperfections is an adequate solution. It may be that the additional eccentricity should be related to the ratio of reinforcement or other influences. Another approach would be to define adequate design values for the modulus of elasticity and the strength, which will result in moment–curvature relationships similar to those shown by the dotted line. But these parameters are already reduced by the safety factors for concrete in contrast to the procedure with steel and it may be that this reduction is already sufficient.

An objective answer to these questions seems to be impossible. Sway members are normally checked against wind loads, the effects of which are much greater than the effects of the imperfections discussed above. In addition, these wind loads are less well known than other design parameters. Therefore, at least in connection with sway structures, the whole question looks to be of theoretical interest rather than of practical use.

2.3 SIMPLIFICATIONS FOR PRACTICAL USE IN SPECIAL CASES

For structures of the same kind, the second order effects are of similar magnitude, so that simple formulae can result in good approximations.

2.3.1 Chimneys

It is economical to design big chimneys with a minimum amount of longitudinal reinforcement. A significant reduction of stiffness then

STABILITY OF COMPRESSION MEMBERS

arises if the cross-sections become cracked. It follows that a good proportion of the shaft should only be cracked very moderately and therefore the deformation behaviour can be described using the stiffness EI of the uncracked cross-section with an adjusting coefficient. Hees (1984) gave the approximation formula in Fig. 2.10, which is valid for cylindrical, conical and parabolic-cylindrical or parabolic-conical shapes. The coefficient of 0·9 was determined on the basis of the German standard DIN 1056 (1984) from a certain number of computer results for typical chimneys, taking care that no underestimation for the additional moment ΔM occurs. The formula is recommended only for additional moments remaining smaller than one-fourth of the first order moment $M.I$ at the base, which is the usual range of moment increase for well designed chimneys. Applying other standards with different safety concepts, for example the CICIND (1980) rules, may need a different coefficient, but the proportionality to the length, the axial force and the stiffness will remain as long as the moment increase is small. The distribution of the additional moment can be described by the given cubic function, also found from computer results, and is independent from the safety concept.

It should be stated that large structures, such as chimneys, are most appropriately designed using computer analysis. However, at this stage, not all designers are familiar with computers and this is why the procedures described above were devised.

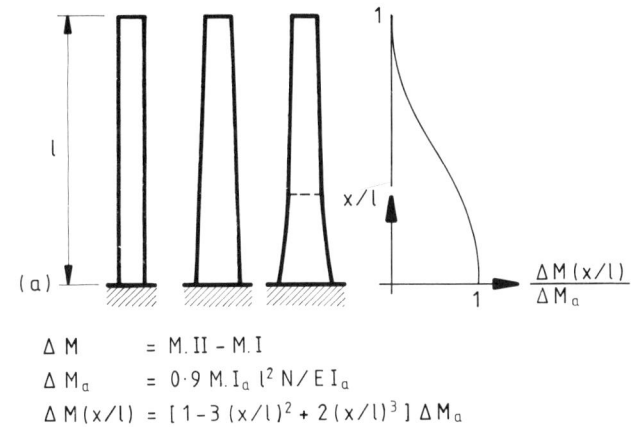

$\Delta M = M.II - M.I$
$\Delta M_a = 0.9 \, M.I_a \, l^2 \, N / E \, I_a$
$\Delta M(x/l) = [1 - 3 \, (x/l)^2 + 2 \, (x/l)^3] \, \Delta M_a$

FIG. 2.10. Simplification of second order effects with chimneys.

2.3.2 Poles and Masts

To make transport and erection easier, poles and masts are designed for minimum weight, leading to high reinforcement ratios of up to 9%. Therefore, in contrast to chimneys, these members behave very much like steel members and in the ultimate limit state yielding of the reinforcement occurs. The tapering of the members is nearly the same, so that the top deflection can be expressed as a function of only the dimensions at the base. When drafting the German standard for poles on the basis of the CEB/FIP Model Code, all types of usual production of lighting columns, transmission poles and antenna masts were examined for sake of calibration of some special coefficients. The result was that the top deflection of such reinforced members in the ultimate limit state can be given by the formula in Fig. 2.11 with a standard deviation of 7·5% compared with the exact result on the basis of the design principles. With a moment increase of 30%, which was found to be the maximum, it is possible to calculate the total actions with an error of less than 3%, which is an adequate result taking into account the uncertainties in defining wind actions. In this case, computer analysis is not justified.

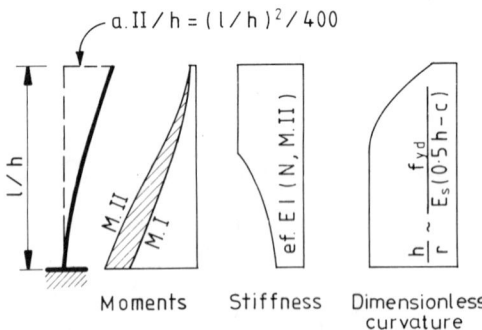

FIG. 2.11. Simplification of second order effects with reinforced concrete poles.

The distribution of the dimensionless curvature h/r in Fig. 2.11 results from the fact that the reinforcement is curtailed in such a way that its strength is always achieved. Calculating the deformations on the basis of stiffnesses is a long way around, because the effective stiffnesses are not known. A similar procedure can be derived for cantilever columns with constant cross-section; the denominator in the formula for deflection is then in the range of 500 to 600 instead of 400.

2.3.3 Sway Frames

Sway frames are defined as structures which suffer a sideways deflection so that second order effects cannot be neglected. The formula at the top in Fig. 2.12 from Eurocode 2 (1984) serves as a criterion for deciding whether a framed structure can be regarded as braced. It takes account of the bracing cross walls or other stiffening elements which behave like cantilever columns. The background of this formula is a very simple one and can be better understood if the formula is squared. The vertical variation of the axial force depends on the number of storeys n. It is constant for a single storey structure but becomes virtually triangular for four and more storeys. The Euler loads for these special distributions are given by the respective expressions. Introducing them in the squared version of the conditional equation shows that the criterion is to keep the ratio of the axial force N of the whole structure and the Euler load such that the moment magnification due to second order effects remains less than 10%. A certain stiffness reduction in the ultimate limit state is also taken into account. Bracing cross walls should remain uncracked under service load conditions.

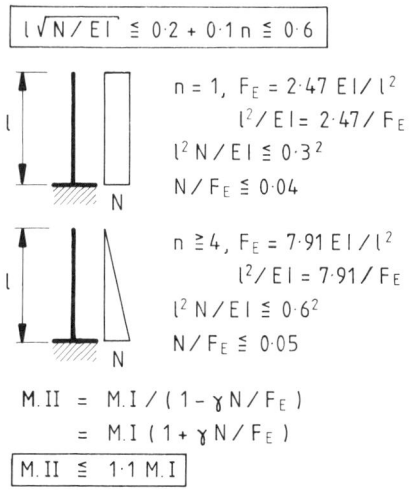

$$l\sqrt{N/EI} \leq 0.2 + 0.1n \leq 0.6$$

$n = 1,\ F_E = 2.47\,EI/l^2$
$l^2/EI = 2.47/F_E$
$l^2 N/EI \leq 0.3^2$
$N/F_E \leq 0.04$

$n \geq 4,\ F_E = 7.91\,EI/l^2$
$l^2/EI = 7.91/F_E$
$l^2 N/EI \leq 0.6^2$
$N/F_E \leq 0.05$

$M.II = M.I/(1 - \gamma N/F_E)$
$\quad\ \ = M.I(1 + \gamma N/F_E)$

$M.II \leq 1.1\,M.I$

FIG. 2.12. Criterion for checking structures against swaying.

Sway frames are checked against stability failure by calculating the action effects of the structure as a whole. Aas-Jakobsen (1973) has reported on a general design method for frames which includes an optimisation for the overall reinforcement. This method is also explained in the CEB/FIP (1978a) *Manual of Buckling and Instability*. For economic reasons the second order effects should always be limited. Then a rough approximation as given in the DAfStb (1979) design manual may suffice. The formula for calculating the total slope from second order theory is shown in Fig. 2.13. The deflected shape of the frame is assumed to be straight with an angle of $\alpha.\mathrm{II}$ from the vertical so that the second order effects may be expressed by the action of additional horizontal forces $\Delta\gamma H = \alpha.\mathrm{II} \times \gamma V$. Assuming uniformly distributed horizontal and vertical loads, the deflection due to these additional forces can be determined by multiplying the first order deflection $a.\mathrm{I}\,(\gamma H)$ due to γ times forces H by the ratio $\alpha.\mathrm{II}\,\Sigma(Vx)/\Sigma(Hx)$. Finally, the total slope can be calculated from the first order top deflection of the frame including the unintended slope ψ. The first order deflections for design actions γH and γV have to be calculated considering the stiffness reduction due to cracking. In many cases a rough estimate is sufficient so that the deflection may be assumed to be twice the deflection belonging to the stiffness EI in the uncracked state. This value is obtained for each frame analysis following first order theory. The

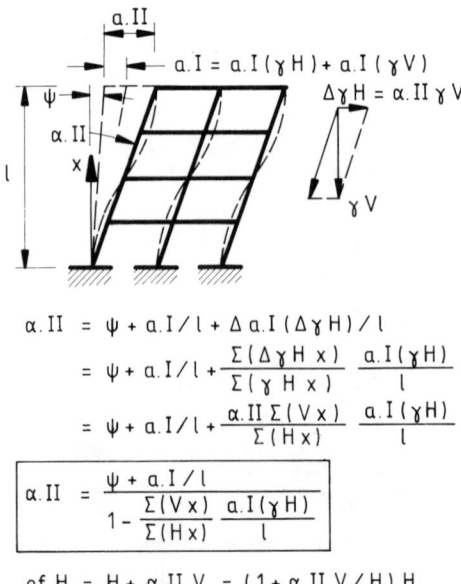

FIG. 2.13. Simplification of the tilt following second order effects.

design of the whole structure then has to be done for increased horizontal loads, as indicated by the last formula in Fig. 2.13. The magnification factor $(1 + a.\mathrm{II}\, V/H)$ for the horizontal loads gives a good indication of the sensitivity of the structure. If it is large, then it may be advisable to modify the design. This simple formula is a special form of the well-known P-Δ-method.

2.3.4 Isolated Columns

The forces throughout braced frames should be calculated from first order theory and thereafter all the columns should be checked, considering them as isolated columns with effective lengths equal to the storey height or not less than 0·85 times that height. Additional second order moments need not be taken into account for checking the restraining members, because sufficient margin is provided with usual constructional design. In contrast to the model column in Fig. 2.2, a reduction of the bearing capacity due to member deflection occurs with columns with uneven end moments only if the critical length, crit.l, is exceeded. This is illustrated in Fig. 2.14. The distribution of the moments or of the total eccentricities — which is the same — of an

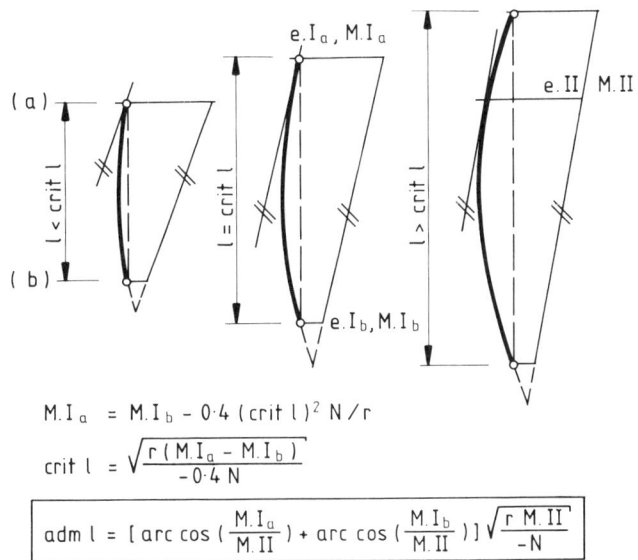

FIG. 2.14. Fixed columns with different end moments.

isolated column with length l = crit.l corresponds to that in a model column with the smaller moment $M.I_b$ at the top and the greater moment $M.I_a$ being the total moment at the base in the deflected state. Substituting in the simplified equation for the model column gives the first equation in Fig. 2.14, which yields the expression for crit.l. This leads to the following design procedure. The isolated column is first designed for the greater first order moment, $M.I_a$. Then it is checked if a greater moment due to deflection results, which is not the case if the calculated critical length is greater than the existing one. A design for an increased second order moment, $M.II$, is only necessary if the existing length is greater than crit.l.

Using the last formula in Fig. 2.14 from Robinson *et al.* (1975), which is better suited for moments with different signs, allows an admissible length to be determined for a chosen value of $M.II$, which must be greater than or equal to the existing length.

Eurocode 2 (1984) contains a simpler design procedure. Checking for stability is only necessary when the effective slenderness exceeds $\lim \lambda = 50 - 25\,(\max M.I/\min M.I)$. Then the design has to be made for an equivalent moment at both ends:

$$\text{equ } M = 0.6 \times \max M.I + 0.4 \times \min M.I > 0.4 \times \max M.I$$

applying for example the model column method or another procedure from CEB/FIP (1978*a*). This procedure is more conservative than the method just described.

2.3.5 Direct Design Method

All the approaches in CEB/FIP (1978*a*) are based on design charts and in addition some do not give the necessary reinforcement directly. Interpolation in design charts often uses as much time as is needed to obtain the necessary reinforcement directly. To increase the number of design charts in order to avoid interpolation is not a practicable solution.

The basic idea of this direct design method is to define fundamental relationships by simple functions, so that additional design charts are not needed and the necessary reinforcement can be calculated. The relationship between the relative axial force n and the relative bending moment m in the ultimate limit state can approximately be defined by simple functions separately for the concrete and steel portions. Figure 2.15 shows the n–m relationship for concrete to be a straight line for $n < -5/6$ and a parabola for $n > -5/6$. For symmetrical reinforcement, concentrated near the edges as shown in Fig. 2.16, n and m depend

STABILITY OF COMPRESSION MEMBERS

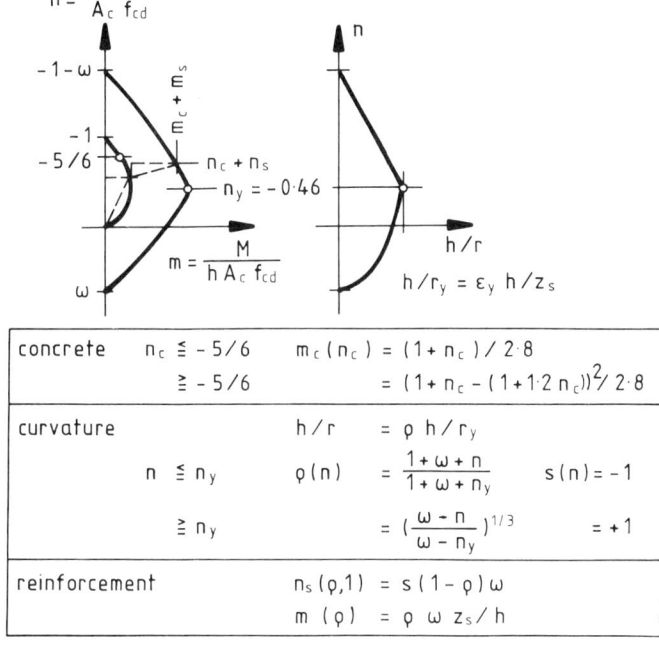

FIG. 2.15. Simplified n–m–(h/r) interaction.

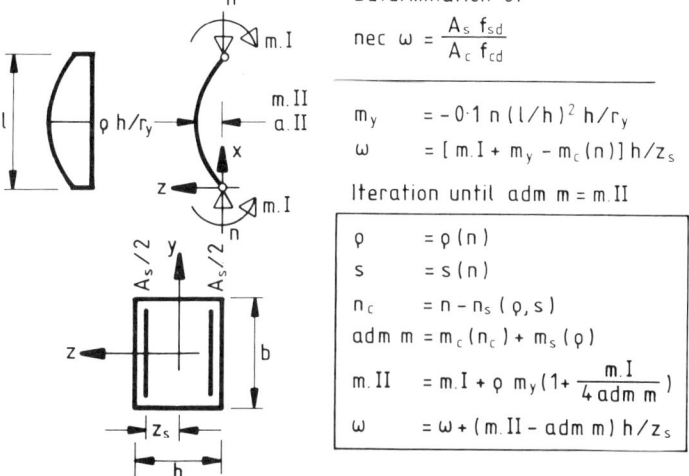

FIG. 2.16. Determination of the necessary reinforcement for a slender column.

linearly on the relative curvature h/r, which can be expressed as a portion ρ of the greatest value, assuming yielding in the reinforcement at the compression and at the tensile side. The interaction between this coefficient ρ and n can be assumed to be a linear one for $n < -0{\cdot}46$ and a cubic one for $n > -0{\cdot}46$. The relative axial force and the bending moment of the steel can be determined for any reinforcement ratio ω. The corresponding formulae and equations are shown in Fig. 2.15 and their usage is explained in the following example, which is example 3.5 from the CEB/FIP (1978a) *Manual of Buckling and Instability* on pages 57 and 58. A column having $l/h = 800/40$ cm $= 20$ with a cross-section $b/h/c = 40/40/4$ cm is to be designed for design actions of $n = -0{\cdot}6$ and $m{\cdot}\text{I} = 0{\cdot}208$ (see Fig. 2.16). The exact solution is $\omega = 0{\cdot}6$ for a steel with a strain at yielding of $\varepsilon_y = 365/200\,000 = 0{\cdot}001\,825$. The calculation follows the formulae in Fig. 2.16.

z/h	=	$(0{\cdot}5 \times 40 - 4)/40$	=	$0{\cdot}4$
m_y	=	$0{\cdot}1 \times 0{\cdot}6 \times 20^2 \times 0{\cdot}001\,825/0{\cdot}4$	=	$0{\cdot}110$
n_c	=	approx. n	=	$-0{\cdot}6$
m_c	=	$(1 - 0{\cdot}6 - (1 - 1{\cdot}2 \times 0{\cdot}6)^2)/2{\cdot}8$	=	$0{\cdot}115$
ω	=	$(0{\cdot}208 + 0{\cdot}110 - 0{\cdot}115)/0{\cdot}4$	=	$0{\cdot}505$
ρ	=	$(1 + 0{\cdot}505 - 0{\cdot}6)/(1 + 0{\cdot}505 - 0{\cdot}46)$	=	$0{\cdot}866$
s	=	-1		
n_s	=	$-1 \times (1 - 0{\cdot}866) \times 0{\cdot}505$	=	$-0{\cdot}068$
n_c	=	$-0{\cdot}6 + 0{\cdot}068$	=	$-0{\cdot}532$
adm.m	=	$(1 - 0{\cdot}532 - (1 - 1{\cdot}2 \times 0{\cdot}532)^2)/2{\cdot}8$		
		$+ 0{\cdot}866 \times 0{\cdot}505 \times 0{\cdot}4$		
		$= 0{\cdot}120 + 0{\cdot}175$	=	$0{\cdot}296$
$m.\text{II}$	=	$0{\cdot}208 + 0{\cdot}866 \times 0{\cdot}110$		
		$\times (1 + 0{\cdot}208/4/0{\cdot}296)$	=	$0{\cdot}319\,(0{\cdot}3187)$
ω	=	$0{\cdot}505 + (0{\cdot}319 - 0{\cdot}296)/0{\cdot}4$	=	$0{\cdot}564\,(0{\cdot}5668)$

The procedure is strictly speaking an iterative one, but the first solution, in general, needs no further iterative improvement. The values in brackets are the final values after three iterations. The deviations in reinforcement area from exact solutions obtained using computers is only a few per cent over the whole range of possible slenderness. The corresponding errors in terms of overall safety are still smaller, in general only a third to a fifth of those of the reinforcement.

More extended formulae can be derived and used with programmable pocket calculators allowing, for instance, columns with unequal end moments to be considered.

STABILITY OF COMPRESSION MEMBERS 67

2.4 PROBLEMS NOT YET COMPLETELY SOLVED

2.4.1 Lateral Buckling, Biaxial Bending

The deflection of columns with rectangular cross-sections in the direction of the smaller side b and with nearly no load eccentricity in this direction is strongly influenced by load eccentricities in the direction of the greater side h, if the ratio is above $h/b > 2$. The reason is that load eccentricities in the direction of h may cause cracking so that a cross-section with a reduced effective width, only comprising the height of the compression zone, is resisting the deflection in the direction of the smaller side b. Habel (1958) reported this problem and proposed to substitute the stiffness reduction by taking an additional eccentricity into account, which leads to a comparable load reduction. This additional eccentricity need only be of the magnitude of the so-called unintended eccentricity. Habel concluded that this problem needs no special consideration, because an unintended eccentricity is always laid down in the design principles. The correctness of this conclusion of course depends on the definition of the nature of the additional eccentricity and whether it is included in an additional overall safety factor, then it certainly cannot be used for compensating shortcomings or simplifications in design procedures, as Habel proposed.

Cranston and Sturrock (1971) reported tests on columns bent about the major axis and buckled sideways. Menegotto and Pinto (1977) and Galgoul (1978) dealt with this problem theoretically. They gave details of the error which arises if the stability check is carried out separately in each principal direction, as allowed by the CEB/FIP (1978b) Model Code in the case that the ratio of the relative eccentricities $(e.b/b)/(e.h/h)$ is smaller than 0·2 or greater than 5·0. They showed that the combination of the load bearing capacities for buckling in the principal directions using the Dunkerley formula can give unsafe results for biaxial buckling. All design methods proposed are complicated, because they do not allow a direct determination of the necessary reinforcement. This problem needs further study to find a practical solution that is economical as well as safe. Eurocode 2 (1984) does not contain any regulations in this aspect.

2.4.2 Slender Walls

Current approaches adopt the results from the linear theory of elasticity for the critical loads of centrically loaded plates, and define effective lengths which are then taken for checking the walls as if they were bars.

Such methods do not take account of factors which mean that eccentrically loaded walls do not behave like eccentrically loaded columns. Acting moments decline due to transversal bending, axial loads may be redistributed from the centre to the stiffened vertical edges and horizontal reinforcement is necessary for ensuring transversal resistance. Research in this area continues and it is hoped that more suitable procedures will be derived before long.

ACKNOWLEDGEMENT

The assistance given by Dr W. B. Cranston of the Cement & Concrete Association, Slough (UK) in drafting this chapter is gratefully acknowledged.

REFERENCES

AAS-JAKOBSEN, K. (1973) *A Design Method for Slender Reinforced Concrete Frames.* Institut für Baustatik, ETH University of Technology, Zürich.

CEB/FIP (1978a) *Manual of Buckling and Instability.* Comité Euro-International du Béton, Bulletin d'Information No. 123.

CEB/FIP (1978b) *Model Code for Concrete Structures.* Comité Euro-International du Béton, Bulletin d'Information No. 124/125-E.

CICIND (1980) *Rules for the Design of Chimneys.* Comité International des Cheminées Industrielles, Düsseldorf.

CRANSTON, W. B. and STURROCK, R. D. (1971) Lateral instability of slender reinforced concrete columns. *Proceedings of the RILEM Symposium, Buenos Aires,* pp. 117–41.

DAFSTB (1979) Bemessung von Beton- und Stahlbetonbauteilen nach DIN 1045, Ausgabe Dezember 1978, 2. überarbeitete Auflage, Schriftenreihe des Deutschen Ausschusses für Stahlbeton (DAfStb), Heft Nr. 220, Wilhelm Ernst & Sohn, Berlin–München–Düsseldorf.

EUROCODE 2 (1984) *Gemeinsame einheitliche Regeln für Beton-, Stahlbeton- und Spannbeton.* Kommission der Europäischen Gemeinschaften, Bericht EUR 8848DE,EN,FR, Entwurf November 1983.

GALGOUL, N. S. (1978) Beitrag zur Bemessung von schlanken Stahlbetonstützen für schiefe Biegung mit Achsdruck unter Kurzzeit- und Dauerbelastung. Dissertation, University of Technology, München. Schriftenreihe des Deutschen Ausschusses für Stahlbeton (DAfStb), Heft Nr. 361, Wilhelm Ernst & Sohn, Berlin–München–Düsseldorf.

GILBERT, R. I. and WARNER, R. F. (1978) Tension stiffening in reinforced concrete slabs. *ASCE, Journal of the Structural Division,* **104**, 1885–900.

HABEL, A. (1958) Knicken senkrecht zur Kraftebene. *Beton- und Stahlbetonbau,* **53**(8), 197–202.

HEES, G. (1984) The development of an approximation formula for the calculation of the bending moments in a concrete chimney shaft. *CICIND, Proceedings of the 5th International Chimney Congress, Essen.*
MENEGOTTO, P. and PINTO, P. E. (1977) Slender RC compressed members in biaxial bending. *ASCE, Journal of the Structural Division,* **103**, 587–605.
NOAKOWSKI, P. and KUPFER, H. (1981) Versteifende Mitwirkung des Betons im Zugbereich von turmartigen Bauwerken, Erfassung und Bedeutung. *Beton- und Stahlbetonbau,* **76**(10, 11), 241–6, 276–80.
QUAST, U. (1977) Auskragender Druckstab — Theorie 2. Ordnung, einachsig, Auskragender Druckstab — Theorie 2. Ordnung, zweiachsig. Anwendungshandbuch zu den Programmen DRUS1 und DRUS2, Deutsche Olivetti GmbH, Frankfurt.
QUAST, U. (1981) Zur Mitwirkung des Betons in der Zugzone. *Beton- und Stahlbetonbau,* **76**(10), 247–50.
QUAST, U. (1984) Zum Stabilitätsverhalten freistehender Stahlbeton-Bauteile. *Beton- und Stahlbetonbau,* **79**(1), 12–16.
RABICH, R. (1969) Beitrag zur Berechnung statisch unbestimmter Tragwerke aus Stahlbeton unter Berücksichtigung der Rißbildung. Aus *Theorie und Praxis des Stahlbetonbaues* (Franz-Festschrift), Wilhelm Ernst & Sohn, Berlin–München–Düsseldorf.
RAO, R. S. (1966) Die Grundlagen zur Berechnung der bei statisch unbestimmten Stahlbetonkonstruktionen im plastischen Bereich auftretenden Umlagerungen der Schnittkräfte. Schriftenreihe des Deutschen Ausschusses für Stahlbeton (DAfStb), Heft 177, Wilhelm Ernst & Sohn, Berlin–München–Düsseldorf.
ROBINSON, FOURÉ and BOURGHLI (1975) Le flambement des poteaux en béton armé chargés avec des excentricités différentes à leurs extrémités. Annales ITBTP, Paris.
SCHLAICH, J., SCHOBER, H. and KOCH, R. (1979) Versuche zur Mitwirkung des Betons in der Zugzone von Stahlbetonröhren. Bericht des Instituts für Massivbau, University of Technology, Stuttgart.

Chapter 3

NONLINEAR ANALYSIS AND OPTIMAL DESIGN OF CONCRETE FRAMED STRUCTURES

C. S. KRISHNAMOORTHY

Department of Civil Engineering, Indian Institute of Technology, Madras, India

SUMMARY

The increasing adoption of limit state design concepts in various Codes of Practice has enhanced the need for nonlinear analysis. Simplified and computer oriented procedures for inelastic analysis are described in this chapter. Nonlinear programming formulations for optimal design of reinforced concrete beam and column sections and frames are presented, and computer software for nonlinear analysis and optimal design are discussed. Further research areas for wider application of these methods are outlined.

NOTATION

A_s	Total area of reinforcement in column section
A_{st}	Area of tension reinforcement
$[A], [\bar{B}]$	Connectivity matrices
$[B]$	Strain-displacement matrix
b	Breadth of the section
C_c	Unit cost of concrete
C_f	Unit cost of formwork
C_s	Unit cost of reinforcement
$[C]$	Constitutive matrix
D	Total depth of the column
$[\bar{D}]$	Diagonal matrix defining the sign of moments
d	Effective depth of the section
d'	Effective cover for reinforcement

E	Modulus of elasticity
$[F]$	Kinematic matrix
f_{ck}	Characteristic strength of concrete
f_{cy}	Cylinder strength of concrete
f_{ij}	Ordinate at j of the influence diagram due to unit value of M_i
f_y	Yield strength of steel
$[G]$	Flexibility matrix for the frame
g_{ij}	Flexibility coefficient $= \int \dfrac{f_i f_j}{EI} ds$ (f_i = influence diagram)
K_1	Ratio of average compressive stress to maximum compressive stress
K_2	Ratio of depth of compression resultant to depth of neutral axis
K_3	Ratio of maximum compressive stress to characteristic strength of concrete
$[K]$	Structure stiffness matrix
$[k]$	Stiffness matrix of all frame elements in block diagonal form
$[k_i]$	Frame element stiffness matrix
L	Length of a member
L_1	Idealised elastic limit
L_2	Idealised plastic limit
M_d	Design bending moment
M_i	Moment at a section i
M_u	Ultimate moment of resistance
M_0	Bending moment at a section in the released structure due to loads
$\{M\}$	Vector of moments
$\{M_c\}$	Vector of moments at cracking
$\{M_e\}$	Elastic moment vector
n	Number of critical sections
P_u	Ultimate axial load
$\{P\}, [L], [N]$	Vector and matrices that are used to define LCP
$\{p\}$	Surface load vector
$\{Q\}$	Nodal load vector
$\{q\}$	Super vector of element vectors
$\{q_i\}$	Element degree of freedom vector
$\{r\}$	Vector of nodal degrees of freedom for the frame/structure

$[S^1], [S^2]$	Matrices to define slopes of first and second segments of moment–rotation relationship
u_{i0}	Angular discontinuity at a basic hinge i due to loads $= \int \dfrac{f_i M_0}{EI} ds$
$\{v\}, \{w\}, \{x\}$	Vectors required to define moment–rotation relationship
$\{X\}$	Body force vector
x_u	Depth of neutral axis
y	Objective function
$[Z]$	Matrix of moments due to unit rotations
α	Total number of critical sections
α'	A scalar quantity
α_s	Shear retention factor
γ_s	Weight density of steel reinforcement
θ_b^i	Inelastic rotation at a basic hinge i
θ_n^j	Inelastic rotation at a non-basic hinge j
θ_p^r	Permissible rotation at a hinge r
θ^r	Inelastic rotation at a hinge r
$\{\theta\}$	Vector of inelastic rotations
$\{\theta_b\}$	Column vector $\{\theta_b^1, \theta_b^2, \ldots, \theta_b^n\}$
$\{\theta_n\}$	Column vector $\{\theta_n^{n+1}, \theta_n^{n+2}, \ldots, \theta_n^\alpha\}$
$\{\theta_y\}$	Vector rotations at yielding

3.1 INTRODUCTION

Reinforced concrete members exhibit inelastic behaviour even at the early stages of loading and this effect is more pronounced at ultimate load stages. Design procedures must ensure adequate safety of the structure under ultimate loads, and the behaviour of the structure under service loads must be satisfactory. Considering these requirements, the CEB–FIP (1970) recommended Limit State Design concepts for reinforced and prestressed concrete structures. Following these recommendations, the Codes of Practice in various countries have been revised and limit state design concepts have been adopted with variations depending on the needs of engineers in each country.

The limit state design method deals with two aspects of design: (a) limit state of collapse and (b) limit states of serviceability. The limit state of collapse is concerned with design of the structure to provide adequate safety against probable overloads and the sections must resist the forces under ultimate load conditions. Limit states of serviceability deal with the checking of deflections and cracking of members under service loads; these must be within the allowable limits for satisfactory behaviour of the structure.

Design for the limit state of collapse requires analysis of structures under ultimate loads. Due to cracking of concrete and inelastic deformations at critical sections, a nonlinear analysis is required to determine the design forces at various sections. Attention is drawn to the interesting papers by Cohn (1964) and Winter (1964), and to the publication edited by Cohn (1972), which discuss the need for nonlinear analysis and design. Recently Ford et al. (1981) highlighted through their detailed experimental investigations the inadequacies of elastic analysis and stressed the need for nonlinear analysis.

The Simplified Limit Method of Baker (1956) and the Imposed Rotations Method of Macchi (1966, 1969) were proposed to provide simplified procedures for nonlinear analysis. These methods were originally oriented to hand computations. However, with increasing use of computers in design practice these methods have been programmed and practical applications have been indicated by Krishnamoorthy (1972), Krishnamoorthy and Yu (1972) and Krishnamoorthy and Mosi (1980). Computer oriented formulations of these methods are presented in Section 3.4.

Finite element analysis offers a powerful numerical tool for rigorous analysis of concrete structures under various stages of loading up to failure considering all the nonlinear effects due to nonlinear stress-strain relations and cracking of concrete, yielding of steel and inelastic deformation of concrete near failure. Nonlinear finite element analysis is still quite expensive and further research work is required to refine the finite element models and to improve the solution algorithms. However, with improvements in computer hardware and wider availability of computing facilities the nonlinear finite element program for analysis of concrete structures may become viable for practical applications. A brief introduction to the finite element analysis is given in Section 3.4.3.

There has been growing interest in the application of mathematical programming techniques to the study of optimising structural design. A survey of work in this area has been presented by Krishnamoorthy and Mosi (1979). Unlike steel structures, the requirements to satisfy the conditions under both ultimate and service loads pose difficulties in mathematical modelling of the reinforced concrete optimal design problem. Cohn and Grierson (1968) and Krishnamoorthy and Munro (1973) linearised the problem and formulated linear programming models for optimal design of frames. The nonlinear optimisation formulations for optimal design of reinforced concrete beam and column sections and frames are described in detail in Section 3.5.

The practical application of nonlinear analysis and optimisation techniques requires availability of computer programs. Further research is also required to study and refine the parameters used in mathematical models. These are highlighted in the concluding section of this chapter.

3.2 MATERIAL CHARACTERISTICS

The realistic determination of the response of reinforced concrete structures requires knowledge of the behaviour of the component materials, concrete and steel, and the ability to incorporate these material characteristics into rational analysis of structures. A brief description of the important properties of concrete and steel is given below.

3.2.1 Concrete under Uniaxial Stress State

Most design calculations are based on the stress–strain characteristics of concrete subject to uniaxial compression. The important factors which affect the response of concrete under load are strength of concrete, lateral reinforcement, creep and shrinkage, strain gradient and specimen size. A large number of investigations were carried out in the 1950s and they have been reviewed by Hognestad *et al.* (1955). The application of a suitable stress–strain relation to the strength calculation of beam and column sections was investigated by Rüsch (1960) through experiments using specially fabricated equipment for controlling the rate of strain. Based on the detailed analysis of various investigations, a simple and practical stress–strain relation for concrete (Fig. 3.1) was suggested by the CEB (1970) for use in the calculations of strength of concrete sections under flexure or combined flexure and axial load. Many Codes of Practice have adopted this simplified parabola-rectangle curve except for a few variations in some of the numerical values.

A general stress–strain relation for concrete under compression

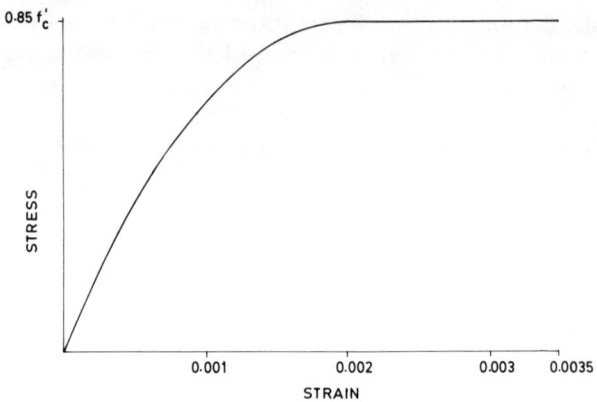

FIG. 3.1. Concrete stress–strain curve (CEB 1970).

which allows for most of the major factors affecting concrete behaviour has been proposed by Sargin (1971). This stress–strain curve shown in Fig. 3.2 can be described by

$$\sigma = K_3 f_{cy} \frac{Ax + (D - 1)x^2}{1 + (A - 2)x + Dx^2} \qquad (3.1a)$$

where A and x are given by

$$A = \frac{E_c \varepsilon_0}{K_3 f_{cy}} \qquad (3.1b)$$

$$x = \frac{\varepsilon}{\varepsilon_0}$$

where ε_0 is the strain corresponding to maximum stress. Sargin (1971) gives details of the values of the above variables.

Though the equation proposed by Sargin (1971) is a generalised one which takes into account most of the factors, determination of the values of A and D is quite difficult. It has been observed that some of the concrete stress–strain relations proposed by various earlier investigators are special cases of equation (3.1a). The parabolic-straight line approximation of the CEB recommendations seems to be a satisfactory idealisation of the uniaxial stress–strain relation in compression that can be used to compute the ultimate strength of sections. However, for complete deformation analysis including cyclic loading, an equation such as eqn (3.1a) is necessary.

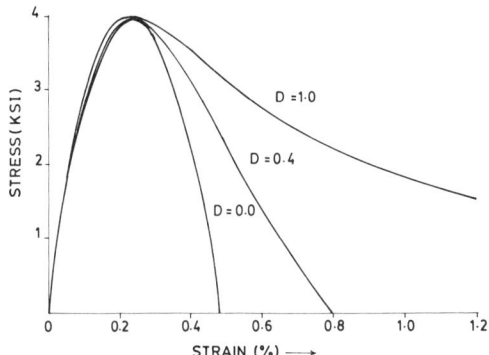

FIG. 3.2. Stress–strain curve (Sargin, 1971).

3.2.2 Concrete under Multiaxial Stress State

A complete nonlinear analysis of reinforced concrete structures under general loading conditions requires knowledge of the generalised material behaviour of concrete under multiaxial stress state. This material behaviour refers to different multiaxial stress–strain relations of concrete when it undergoes physical changes like cracking in tension, yielding in compression and also failure of concrete. Strength of concrete can be determined only by considering the interaction of various components of the states of stress.

The experimental investigations of concrete subject to multiaxial stress state are limited to support any rational formulation of constitutive relations. Chen (1982) has presented in detail the following theories for constitutive modelling of concrete: (a) linear elasticity theory, (b) nonlinear elasticity theory, (c) perfect and work hardening plasticity theories, and (d) endochronic theory of plasticity. While linear elasticity theory can be used for concrete subjected to tension and very low compression, the other theories can be used to study the behaviour of concrete under increasing compressive stresses up to failure. A study of these theories is useful for nonlinear finite element analysis of concrete structures.

3.2.3 Stress–Strain Relation for Steel

Figure 3.3 shows the tensile stress–strain curves of typical mild steel and cold-worked steel bar. In many of the simplified analysis procedures, assumption of bilinear, elastic–perfectly plastic behaviour has been found to be adequate.

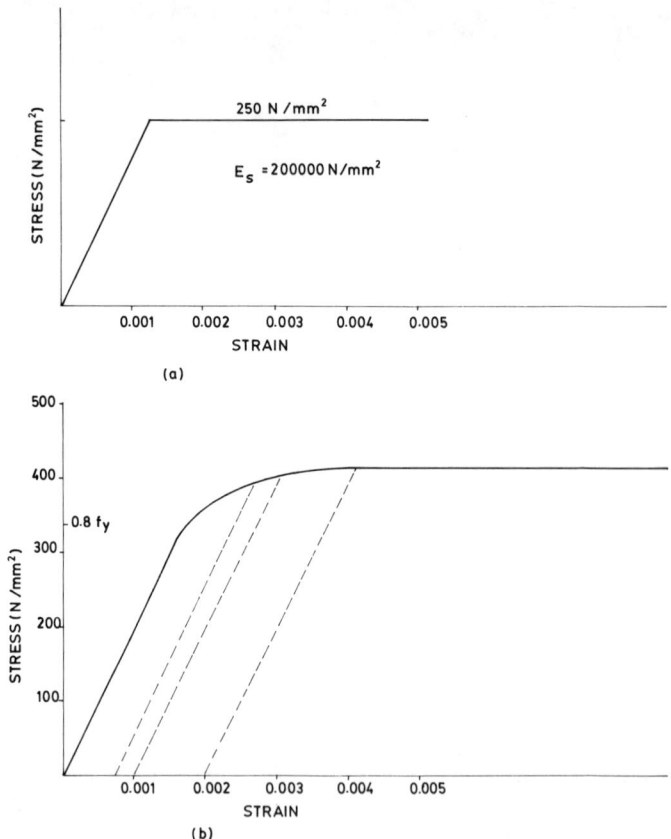

FIG. 3.3. Stress–strain curves for steel. (a) Mild steel; (b) cold-worked steel.

3.3 DEFORMATION CHARACTERISTICS OF REINFORCED CONCRETE MEMBERS

The behaviour of reinforced concrete members is best represented by the moment–curvature relations of the cross-sections. Figure 3.4(a) shows the moment–curvature relation for a cross-section of a beam member with a normal or small percentage of reinforcement. In this case three phases can be observed:

Phase 1: the straight line segment OC representing linear behaviour as the concrete has not yet cracked.

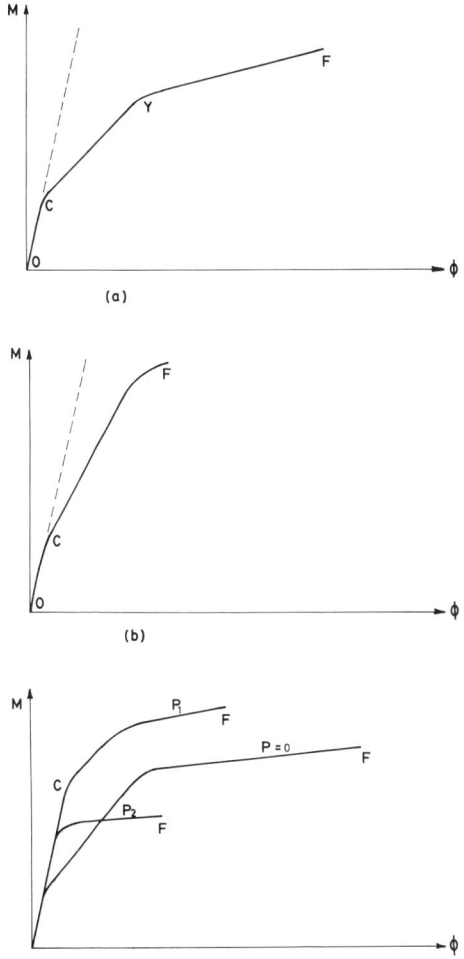

FIG. 3.4. Moment (M)–curvature (ϕ) relations for reinforced concrete sections. (a) Under-reinforced; (b) over-reinforced; (c) bending and axial load.

Phase 2: the segment between cracking of concrete (C) and yielding of steel (Y).
Phase 3: after yielding of steel (Y), inelastic rotations set in resulting in large concrete strain. Failure occurs when concrete reaches the maximum strain, indicated by the point F.

Figure 3.4(b) shows the relation for an over-reinforced section; here failure occurs before the yielding of steel and the deformations are due to inelastic behaviour of the concrete.

For members subjected to axial load and bending, the moment–curvature relations show similar behaviour (Fig. 3.4(c)) except that the ductility is much more limited due to the presence of the axial load. Failure can also occur without cracking for small eccentricities of the axial load (P_2) when the entire section is under compression.

3.3.1 Moment–Rotation Characteristics

If the moment–curvature relations of sections are known, it is theoretically possible to analyse the structure for any loading up to and including the ultimate load. But this approach, using moment–curvature relations, leads to complicated and cumbersome calculations due to the nonlinear nature of the relationship. It will be shown in Section 3.4 that the inelastic analysis of framed structures requires identification of critical sections of a member which undergo large inelastic deformation. It is, therefore, logical to use the moment–rotation relation for a critical section and thus avoid the disadvantages inherent in using the moment–curvature relation. The inelastic rotation is not a sectional property but is the cumulative effect of the behaviour of the reinforced concrete member in the vicinity of the critical section; it is not a discontinuous effect. This concept of moment–rotation relation has been used by Baker (1956, 1970) and Macchi (1966) and found to be convenient for analysis of frames under ultimate load. Extensive experimental investigations carried out on the moment–rotation characteristics of beams and columns are reported by Baker and Amarakone (1964) and Macchi and Siviero (1974). Typical moment–rotation relations are shown in Fig. 3.5.

It can be seen from Fig. 3.5 that the moment–rotation curve is non-linear even at the early stages of loading. The slope of the curve at the origin is influenced by the uncracked flexural rigidity of the member. As the load increases, the flexural rigidity decreases and does not have a constant value throughout the member. After yield the rotation at a critical section of the member comprises (a) the contribution from that portion of the member acting elastically, and (b) the contribution from that portion of the member in the vicinity of the critical section undergoing inelastic deformation.

3.3.2 Flexural Rigidity of Reinforced Concrete Members

For the analysis of reinforced concrete framed structures, the flexural

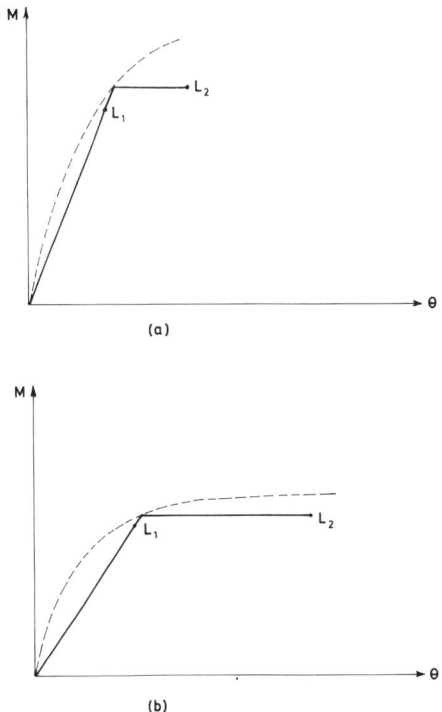

FIG. 3.5. Typical moment (M)-rotation (θ) relationships. (a) Column; (b) beam.

rigidity EI of members forms an important factor influencing the calculated stress state. For the elastic analysis of these structures, the Codes of Practice suggest that the relative stiffnesses may be calculated using any one of the following throughout:

(a) the concrete section — the entire concrete cross-section ignoring the reinforcement;
(b) the gross section — the entire concrete cross-section including the reinforcement on the basis of modular ratio;
(c) the transformed section — the compression area of the concrete cross-section combined with the reinforcement on the basis of modular ratio.

The difference in the EI values calculated by different assumptions is quite large. In the analysis of frames the basic assumptions for the determination of EI values may give rise to large differences in the design values. For certain frames under extreme conditions the error in the analysis of moments due to different assumptions for the EI values could be as much as 30%.

Experimental investigations show that there is considerable reduction in stiffness of reinforced concrete members at the ultimate load stage. This is also evident from the moment–curvature diagrams (Fig. 3.4). In the analysis of frames for various limit states it is important to base the calculation of EI values consistent with the behaviour of the members at the relevant load stage. The Codes of Practice recommend different expressions based on experimental investigations that can be used to compute deflections under service loads, i.e. to check limit state of serviceability. Baker (1956) and Macchi (1966) have proposed procedures for analysis under limit state of collapse and these are discussed in Section 3.4.

3.3.3 Inelastic Rotation Capacity of Reinforced Concrete Members

Analysis of reinforced concrete framed structures for limit state of collapse requires knowledge of the rotation capacity of hinging regions of the members. The limit of the plastic rotation is primarily controlled by the ultimate strain in concrete or in extremely under-reinforced members the strain in the reinforcement may determine the maximum plastic rotation.

Extensive experimental investigations on reinforced concrete members by Baker and Amarakone (1964) show that the plastic rotation capacity $\theta_p{}^r$ is subject to large fluctuations even when the members are tested under similar conditions. Hence, in calculations involving plastic rotations, it would only be possible to use approximate values of $\theta_p{}^r$, which may be considered to be safe values as compared to experimental results. Baker and Amarakone (1964) proposed a set of curves shown in Fig. 3.6 for the calculation of rotation capacity of reinforced concrete members. This set of curves has been found suitable for use in analysis and design as it involves a minimum number of variables.

3.4 NONLINEAR ANALYSIS OF REINFORCED CONCRETE FRAMED STRUCTURES

It is evident from study of the deformational characteristics described in the previous section that the behaviour of reinforced concrete members is nonlinear due to the nonlinear stress–strain relation of concrete, cracking of concrete and yielding of steel and the inelastic deformation

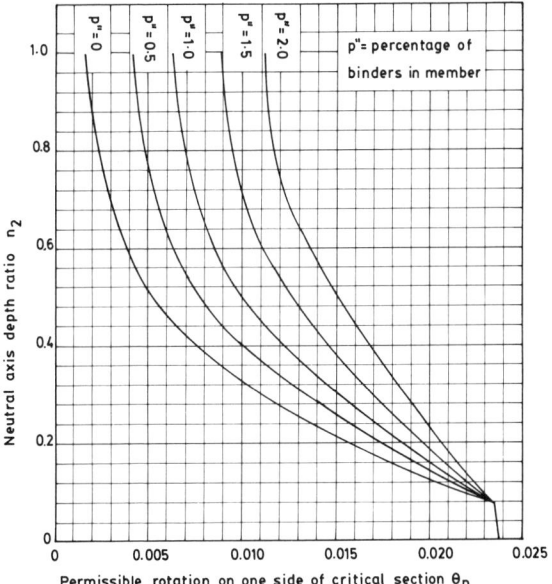

FIG. 3.6. Idealised plastic rotation capacity of reinforced concrete members.

of concrete at ultimate load stages. Hence any rigorous analysis of reinforced concrete structures should consider these nonlinear effects, which is now possible through use of the finite element technique. However, simplified analysis procedures have been proposed by Baker (1956) and Macchi (1966), which provide powerful tools for analysis of frames under limit state of collapse. These simplified methods of inelastic analysis and rigorous nonlinear finite element analysis techniques will be described in this section.

3.4.1 Simplified Limit Method

Baker (1956) proposed a simplified procedure for analysis of frames in the ultimate condition. The assumptions made in this method are as follows:

1. The moment–rotation diagram of reinforced concrete members is idealised to be bilinear as shown in Fig. 3.5. The flexural rigidity EI of the members is, therefore, idealised to be constant from 0 to L_1, and between the critical sections of the member.

   The limits L_1 and L_2 are so chosen that they are on the safe side yet as close to the actual diagram as possible. The limit L_1 can be

termed the idealised elastic limit and L_2 the idealised plastic limit. These limits are defined by specifying the stress–strain relation for concrete and the limiting strain conditions at the critical section. Experimental evidence shows that the idealised elastic limit L_1 is associated with the yielding of tension reinforcement or the concrete reaching maximum stress. Baker (1970) has suggested that the limit L_1 can be defined by the occurrence of two conditions, viz. the attainment of a strain of 0·002 in concrete or an offset strain of 0·001 in steel, whichever is attained first. The limit L_2 is defined by the attainment of a maximum strain of 0·0035 for unbound concrete or 0·01 in steel.
2. The inelastic rotation is concentrated at a critical section which acts as a plastic hinge.
3. The ultimate strength of a frame is reached when a sufficient number of critical sections become plastic hinges to make the structure statically determinate and stable.
4. Members between plastic hinges are assumed to behave elastically and are designed to resist bending moments equal to or less than the values at limit L_1 for the members.
5. The positions of the first n plastic hinges are predetermined and as implied in assumption (4) there is no plastic rotation at any other position.

The method is essentially based on the flexibility method of structural analysis. For a given frame indeterminate to the nth degree, n hinge releases are chosen to make it determinate and stable. These hinges are chosen from a knowledge of approximate elastic moment distribution, at positions where the bending moments are critical for the members. Figure 3.7 shows a suggested pattern of hinges for a frame subjected to wind, dead and live loads.

The governing compatibility equation in terms of the redundant moments M_i acting at the hinge locations i is expressed as follows:

$$g_{11}M_1 + g_{12}M_2 + \ldots + g_{1n}M_n + u_{10} = -\theta_b^1$$
$$g_{21}M_1 + g_{22}M_2 + \ldots + g_{2n}M_n + u_{20} = -\theta_b^2$$
$$\ldots \qquad \ldots \qquad \ldots \qquad \qquad (3.2)$$
$$g_{n1}M_1 + g_{n2}M_2 + \ldots + g_{nn}M_n + u_{n0} = -\theta_b^n$$

NONLINEAR ANALYSIS AND OPTIMAL DESIGN 85

The above set of equations (3.2) are solved for moments M_i by a trial and error adjustment procedure briefly described below:

1. Work out the coefficients g_{ij} and u_{i0} for the chosen release system of the type shown in Fig. 3.7.
2. Choose a set of M_i values and adjust their values by a systematic procedure such that the rotations θ_b^i at these hinges are less than the permissible values.

The guidelines for the initial choice of the values of M_i and the method of working the trial and adjustment procedure are fully explained by Baker (1956).

For the analysis of large frames the above method poses problems of convergence, and a computer oriented procedure has been suggested by

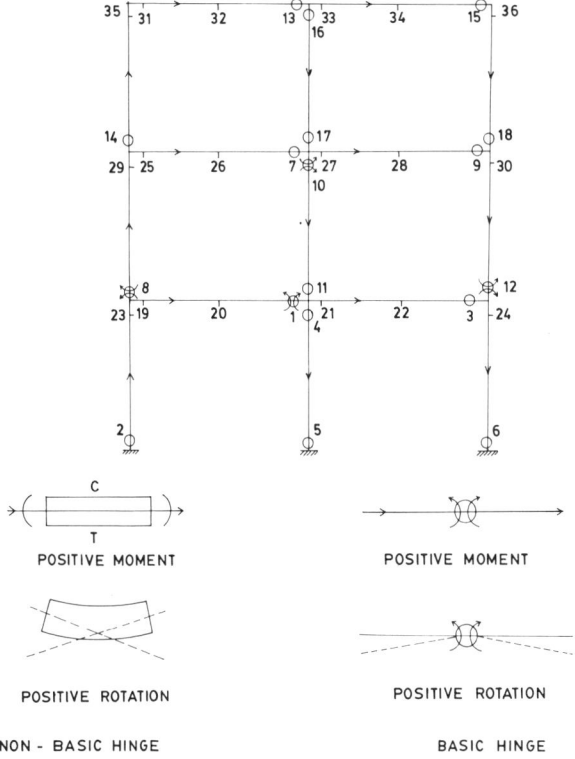

FIG. 3.7. Hinge system for a frame, and sign conventions.

Krishnamoorthy and Yu (1972). For the development of such a computer program it has been found desirable to base the analysis on the generalised compatibility method proposed by Munro (1965).

3.4.1.1 *General Limit Method*

A general limit analysis method using a set of influence functions for a released structure was proposed by Munro (1965), who described a suitable procedure for computation of collapse loads and displacements.

A frame which is n times indeterminate can be reduced to a determinate structure by suitable insertion of hinges, as shown for a typical frame in Fig. 3.8. The selected hinges are termed 'basic' and are

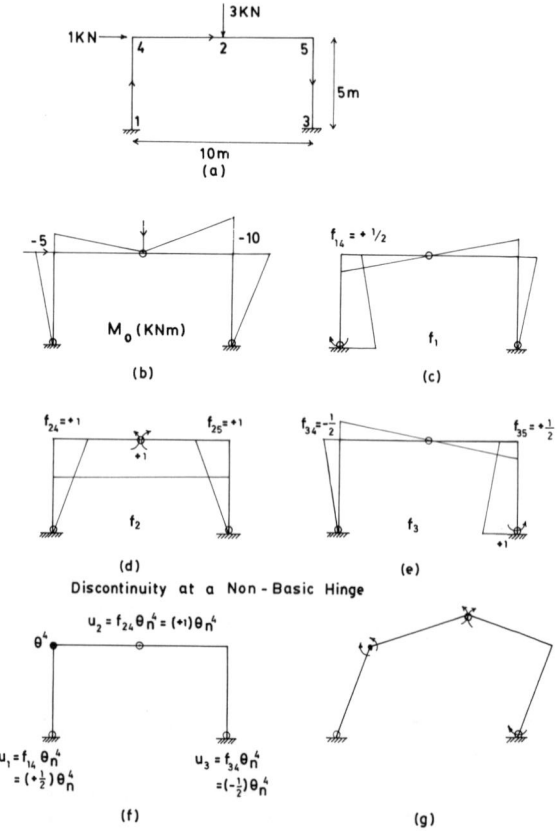

FIG. 3.8. Influence diagrams for a frame.

numbered 1 to n. The remaining critical sections are termed 'non-basic' and are numbered $n + 1$ to a where a is the total number of critical sections.

The basic influence diagrams f_i, which are bending moment diagrams due to unit values of moments applied at the basic hinges, and the moment diagram for a particular loading condition M_0 are shown for a typical frame in Fig. 3.8. It can be shown by the principle of contragredience that the influence diagrams can be used for the computation of discontinuities caused at the basic hinges due to inelastic rotation at non-basic hinges. Thus, the angular discontinuity at a basic hinge 1 for the frame shown in Fig. 3.8 is given by

$$\theta_b^1 = g_{11}M_1 + g_{12}M_2 + g_{13}M_3 + u_{10} + f_{14}\theta_n^4 + f_{15}\theta_n^5 \qquad (3.3)$$

where θ_n^4 and θ_n^5 are the inelastic rotations at the non-basic hinge sections 4 and 5, and f_{14} and f_{15} are the ordinates of the influence diagrams as shown in Fig. 3.8.

Similarly, the angular discontinuity at other hinges can be expressed and the compatibility equation for the frame becomes

$$g_{11}M_1 + g_{12}M_2 + g_{13}M_3 + u_{10} + f_{14}\theta_n^4 + f_{15}\theta_n^5 = \theta_b^1$$
$$g_{21}M_1 + g_{22}M_2 + g_{23}M_3 + u_{20} + f_{24}\theta_n^4 + f_{25}\theta_n^5 = \theta_b^2 \qquad (3.4)$$
$$g_{31}M_1 + g_{32}M_2 + g_{33}M_3 + u_{30} + f_{34}\theta_n^4 + f_{35}\theta_n^5 = \theta_b^3$$

The above result can be extended to a frame indeterminate to the nth degree and the compatibility equation is as follows:

$$g_{11}M_1 + g_{12}M_2 + \ldots + g_{1n}M_n + u_{10} + \sum_{j=n+1}^{a} f_{1j}\theta_n^j = \theta_b^1$$

$$g_{21}M_1 + g_{22}M_2 + \ldots + g_{2n}M_n + u_{20} + \sum_{j=n+1}^{a} f_{2j}\theta_n^j = \theta_b^2$$

$$\ldots$$

$$\ldots$$

$$g_{n1}M_1 + g_{n2}M_2 + \ldots + g_{nn}M_n + u_{n0} + \sum_{j=n+1}^{a} f_{nj}\theta_n^j = \theta_b^n \qquad (3.5)$$

and expressed in matrix form

$$[G]\{M\} + \{u_0\} + [F]\{\theta_n\} = \{\theta_b\} \qquad (3.6)$$

Equation (3.6) can be solved for $\{M\}$ and is given by

$$\{M\} = [G]^{-1}\{\{\theta_b\} - [F]\{\theta_n\} - \{u_0\}\} \qquad (3.7)$$

Parity rule. In the ultimate load analysis of frames with some of the sections developing inelastic rotations, it is important to establish a parity rule between the moments and rotations at these critical sections. The sign conventions used for arbitrarily selected member directions are shown in Fig. 3.7. In order that the stress resultants correspond to the strain resultants, the following parity rule proposed by Munro (1965) must be satisfied.

At a basic hinge, i developing inelastic rotation:

$$\text{when} \quad M_i \text{ is positive}, \ \theta^i_b \leq 0$$
$$\text{and when} \quad M_i \text{ is negative}, \ \theta^i_b \geq 0 \qquad (3.8a)$$

At a non-basic hinge j developing inelastic rotation:

$$\text{when} \quad M_j \text{ is positive}, \ \theta^j_n \geq 0$$
$$\text{and when} \quad M_j \text{ is negative}, \ \theta^j_n \leq 0 \qquad (3.8b)$$

3.4.1.2 *Modified Procedure by Adjustment of Rotations*

A modified procedure to the simplified limit method has been proposed by Krishnamoorthy and Yu (1972); this procedure has been found to be suitable for computer programming and provides a rapid solution. The procedure is based on the adjustment of rotations in an iterative manner and is described below.

1. Assign the permissible rotation θ_p^r, as the value of θ^r for beam hinges whose sign of moments are known for normal cases and zero rotation for column hinges. With these values of θ^r analyse the frame using eqn (3.7) for moments M_i at all the hinges. The solution gives the magnitude and sign of moments at all the critical sections.

2. Since the sign of the column moments are now known, assign the sign of θ_p^r for column hinges according to the parity rule (eqn 3.8). Analyse the frame with proper values of θ_p^r as θ^r for beam and column hinges. The moments at the critical beam and column sections corresponding to the permissible hinge rotations are now determined.

3. (a) Compare the moments M_r obtained and the sign θ^r used for the analysis in the previous step, and thus check for the parity rule.

3. (b) If the parity rule is not satisfied, assign the value of θ^r at that hinge to be zero. Analyse the frame with the values of θ^r as zero at those

hinges where the parity rule is not satisfied and θ_p^r as the values of θ^r at the other hinges.

4. Check the parity rule for the analysis carried out in step (3). If it is not satisfied at all the hinges, repeat the procedure described in step 3(b) until it is satisfied at all the hinges.

This procedure has been found to be quite effective for practical application to regular multistorey frames.

3.4.2 Mathematical Programming Model

The mathematical programming model is based on the method of imposed rotations and is due to De Donato and Maier (1972) and Maier *et al.* (1972).

The stiffness matrix method of analysis is adopted for the elastic analysis of the frame. Let the inelastic rotations at the n critical sections be denoted by the vector $\{\theta\}$. The critical sections are always located adjacent to nodes where the elements of the frame are connected to each other. The stiffness matrix $[k_i]$ of the frame element is calculated using the well known expressions. If $\{q\}$ represents the super vector of each element degree of freedom $\{q_i\}$, as shown in Fig. 3.9, the compatibility condition for the frame can be expressed as

$$\{q\} = [A]\{r\} + [\bar{B}]\{\theta\} \tag{3.9}$$

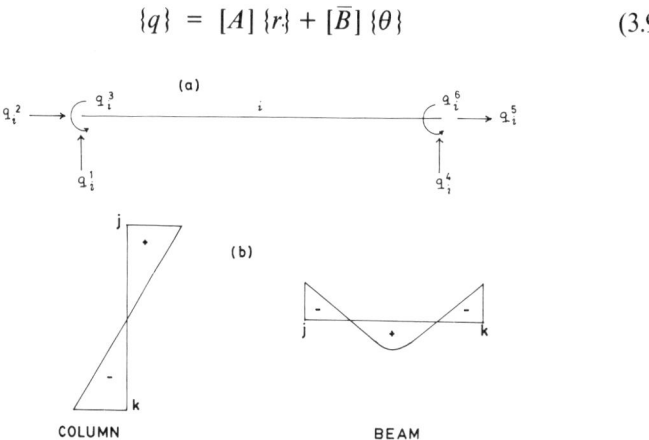

FIG. 3.9. Frame element. (a) degrees of freedom, (b) sign conventions for moments.

where $[A]$ and $[B]$ are connectivity matrices and $\{r\}$ is the vector of **nodal** degrees of freedom for the frame.

If the nodal load vector is represented by $\{Q\}$, the equilibrium for the frame is given by

$$[k]\{q\} = \{Q\} \tag{3.10}$$

where $[k]$ is the stiffness matrix of all elements in block diagonal form. Substituting eqn (3.9) in eqn (3.10), we can express the equilibrium equation as

$$[K]\{r\} = \{\bar{Q}\} - [A]^T [k] [\bar{B}] \{\theta\} \tag{3.11}$$

where $[K]$ is the structure stiffness matrix $= [A]^T [k] [A]$ and $\{\bar{Q}\} = [A]^T \{Q\}$.

Equation (3.11) when solved with respect to $\{r\}$ gives the nodal displacements due to external loads and due to rotations $\{\theta\}$. The elastic solution of the frame can be obtained by assigning $\{\theta\} = \{0\}$. Similarly, the response due to imposed rotations $\{\theta\}$ can be obtained by putting $\{\bar{Q}\} = \{0\}$. Thus

$$\{r\} = -[K]^{-1} [A]^T [k] [\bar{B}] \{\theta\} \tag{3.12}$$

The elastic moment vector is given by

$$\{M_e\} = [\bar{B}]^T \{Q\} \tag{3.13}$$

After eliminating for $\{q\}$ and $\{r\}$ and simplifying, eqn (3.13) reduces to

$$\{M_e\} = [\bar{B}]^T [-[k] [A] [K]^{-1} [A]^T [k] [\bar{B}] + [k] [\bar{B}]] \{\theta\} \tag{3.14}$$

Thus, the moment at critical sections due to unit values of $\{\theta\}$ can be expressed by the matrix $[Z]$ as

$$[Z] = [\bar{B}]^T [-[k] [A] [K]^{-1} [A]^T [k] + [k]] [\bar{B}] \tag{3.15}$$

The idealised moment–rotation relation to be used in the analysis is shown in Fig. 3.10. Mathematically this relation can be expressed by the following equations:

$$M = \bar{d} M_c + \bar{d} S^1 v + \bar{d} S^2 x - \bar{d} w \tag{3.16}$$

$$w, v, x \geqslant 0, v \leqslant \theta_y \tag{3.17}$$

$$wv = 0, (\theta_y - v)x = 0 \tag{3.18}$$

$$\theta = \bar{d}(v + x) \tag{3.19}$$

where $\bar{d}$ is 1 if the section is positive or -1 if the section is negative. Similarly, the moment–rotation relationship can be expressed for all the n critical sections, and in a compact form is given by

$$\{M\} = [\bar{D}]\{M_c\} + [\bar{D}][S^1]\{v\} + [\bar{D}][S^2]\{x\} - [\bar{D}]\{w\} \tag{3.20}$$

$$\{w\}, \{v\}, \{x\} \geq \{0\}, \{v\} \leq \{\theta_y\} \quad (3.21)$$
$$\{w\}^T \{v\} = 0, \quad [\{\theta_y\} - \{v\}]^T \{x\} = 0 \quad (3.22)$$
$$\{\theta\} = [\bar{D}] [\{v\} + \{x\}] \quad (3.23)$$

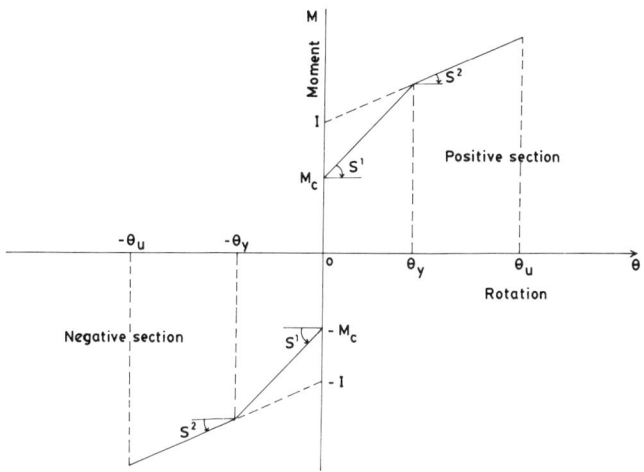

FIG. 3.10. Idealised trilinear M-θ relation.

By the principle of superposition, the moment at critical sections can be expressed as
$$\{M\} = a' \{M_e\} + [Z] \{\theta\} \quad (3.24)$$
where $0 \leq a' \leq 1$.

Substituting from eqns (3.20) and (3.23) into eqn (3.24) gives
$$[\bar{D}] [M_c] + [\bar{D}] [S^1] \{v\} + [\bar{D}] [S^2] \{x\} - [\bar{D}] \{w\}$$
$$= a' \{M_e\} + [Z] [\bar{D}] [\{v\} + \{x\}] \quad (3.25)$$

Multiplying eqn (3.25) by $[\bar{D}]^{-1}$ and rearranging the terms gives
$$\{w\} = \{M_c\} + a' \{P\} + [L] \{v\} + [N] \{x\} \quad (3.26)$$
where
$$\{P\} = -[\bar{D}] \{M_e\}$$
$$[L] = [S^1] - [\bar{D}] [Z] [\bar{D}]$$
$$[N] = [S^2] - [\bar{D}] [Z] [\bar{D}]$$

The solution for eqn (3.26) is subject to the conditions to be satisfied as given by eqns (3.21) and (3.22).

In addition, to ensure that

$$[\bar{D}]\{M\} \geqslant \{0\} \quad \text{if} \quad \{\theta\} = \{0\}$$

the following condition must also be satisfied:

$$\{w\} \leqslant \{M_c\} \tag{3.27}$$

The mathematical model defined by eqns (3.26), (3.22), (3.23) and (3.27) is called a $n \times 2n$ Linear Complementarity Problem (LCP). This problem has a unique solution $\{w\}, \{v\}, \{x\}$, that is when $\alpha' = 1$ such that $\{M\}$ and $\{\theta\}$ given by eqns (3.24) and (3.23) satisfy the moment–rotation relation. The solution of the above problem with practical application has been reported by De Donato and Maier (1972), Maier et al. (1972) and Kaneko (1977). The computational aspects with program flow charts have been illustrated by Krishnamoorthy and Mosi (1980).

3.4.3 Finite Element Analysis of Concrete Structures

The finite element method provides a powerful numerical technique for rigorous analysis of structures of complex material properties and complicated boundary conditions. With the increasing availability of large computing facilities, this method has been used by many investigators to study the nonlinear behaviour of concrete structures. The application to reinforced concrete beams was first attempted by Ngo and Scordelis (1967). Scordelis (1972), in his state-of-the-art report, reviewed the status of research carried out up to 1971 and cited several applications. Colville and Abbasi (1974) considered the reinforcement as an integral part of the concrete element, thus providing facility for discretisation independent of the reinforcement position. This study was followed by Panneerselvam (1977) and Krishnamoorthy and Panneerselvam (1977, 1978), who used a four-noded isoparametric quadrilateral element with incompatible modes and successfully applied the method to study the nonlinear behaviour of reinforced concrete framed structures. The finite element formulation is briefly explained below.

Consider a reinforced concrete element with steel reinforcement as shown in Fig. 3.11. The total potential energy for the element can be written as

$$\pi = U_c + U_s - U_{cc} - U_{yc} - U_{ys} + W \tag{3.28}$$

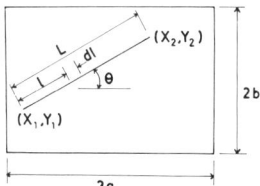

FIG. 3.11. A steel reinforcement in the element.

where U_c is the strain energy of the elastic uncracked concrete, U_s is the strain energy of unyielded steel, U_{cc} is the strain energy released due to cracking of concrete, U_{yc} is the strain energy released due to yielding of concrete, U_{ys} is the strain energy released due to yielding of steel and W is the potential energy due to external loads.

For finite element displacement formulation, the total potential energy of an element is given by

$$\pi = \frac{1}{2}\{q_i\}^T \left(\int_{V_c} [B]^T [C_c] [B] \, dV_c + \int_{V_s} [B]^T [C_s] [B] \, dV_s \right) \{q_i\}$$

$$- \frac{1}{2}\{q_i\}^T \left(\int_{V_{cc}} [B]^T [C_{cc}] [B] \, dV_{cc} + \int_{V_{yc}} [B]^T [C_{yc}] [B] \, dV_{yc} \right.$$

$$\left. + \int_{V_{ys}} [B]^T [C_{ys}] [B] \, dV_{ys} \right) \{q_i\}$$

$$- \{q_i\}^T \{Q_n\} - \{q_i\}^T \int_V [N]^T \{X\} \, dV$$

$$- \{q_i\}^T \int [N]^T \{p\} \, dS \tag{3.29}$$

where $[C_c]$ is the material constitutive matrix for elastic uncracked concrete, $[C_s]$ is the material constitutive matrix for unyielded steel, $[C_{cc}]$ is the change in material constitutive matrix due to cracking of concrete, $[C_{yc}]$ is the change in material constitutive matrix due to yielding of concrete, $[C_{ys}]$ is the change in material constitutive matrix due to yielding of steel.

From the above relations, the stiffness matrix of an elastic uncracked reinforced concrete element can be expressed as

$$[k_{cs}] = [k_c] + [k_s] \tag{3.30a}$$

where

$[k_c]$ = stiffness matrix of concrete = $\dfrac{1}{2} \iiint [B]^T [C_c] [B] \, dV_c$ (3.30b)

$$[k_s] = \sum_{i=1}^{n_s} A_i \int [B]^T [C_s]_i [B] \, dl \qquad (3.30c)$$

and n_s is the total number of bars in the element.

By assembling all the elements, the overall equation of equilibrium of the structure is expressed as

$$[K]\{r\} = \{R\} \qquad (3.31)$$

The nonlinearities included in the formulation are the stress–strain relation, tensile cracking, and yielding of concrete and yielding of steel reinforcement. These effects are considered through the change in the constitutive relations as briefly indicated below.

Among the various uniaxial stress–strain relations available, the one adopted in the CEB recommendation (discussed in Section 3.2) has been found to be adequate for monotonic loading. Actually the concrete is in a state of biaxial stress. However, a simple state of isotropic condition with a minimum value of the elastic modulus E for concrete in biaxial compression was found to be adequate, and accordingly the constitutive matrix $[C]$ is given by

$$[C] = \dfrac{E}{1-\mu^2} \begin{bmatrix} 1 & \mu & 0 \\ \mu & 1 & 0 \\ 0 & 0 & \dfrac{(1-\mu)}{2} \end{bmatrix} \qquad (3.32)$$

where μ is Poisson's ratio.

The yield and failure criteria due to Kupfer *et al.* (1969) for plain concrete in biaxial compression based on their experimental investigation can be used in the analysis.

When one of the principal stresses exceeds the tensile strength, concrete is considered to have cracked in a direction perpendicular to that principal stress. The normal stress at the crack drops to zero. The shear modulus is also reduced by cracking. Referring to Fig. 3.12, $X'Y'$

is the coordinate system parallel and perpendicular to the crack. The stress-strain relation in the $X'Y'$ coordinate system is given by

$$\begin{Bmatrix} \sigma'_x \\ \sigma'_y \\ \tau'_{xy} \end{Bmatrix} = \begin{bmatrix} E & 0 & 0 \\ 0 & 0 & 0 \\ 0 & 0 & \alpha_s G \end{bmatrix} \begin{Bmatrix} \varepsilon'_x \\ \varepsilon'_y \\ \gamma'_{xy} \end{Bmatrix} \qquad (3.33)$$

where $G = \dfrac{E}{2(1+\mu)}$ and α_s is the cracked shear constant.

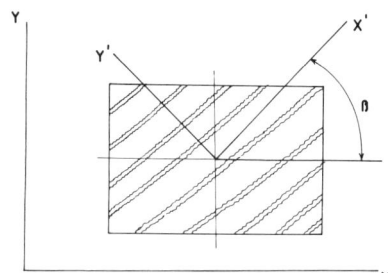

FIG. 3.12. Cracked element.

If both the principal stresses exceed the tensile strength, the concrete will crack in both the principal stress directions. Again the stresses in the cracked element can be written

$$\begin{Bmatrix} \sigma'_x \\ \sigma'_y \\ \tau'_{xy} \end{Bmatrix} = \begin{bmatrix} 0 & 0 & 0 \\ 0 & 0 & 0 \\ 0 & 0 & \alpha_s G \end{bmatrix} \begin{Bmatrix} \varepsilon'_x \\ \varepsilon'_y \\ \gamma'_{xy} \end{Bmatrix} \qquad (3.34)$$

The shear retention factor α_s introduced here will retain a certain amount of shear stress in the cracked concrete.

The steel reinforcement is subject to a uniaxial stress state and the stress–strain relation including yielding can be considered.

The nonlinear finite element equations are solved by using incremental iterative procedures for various stages of loading up to failure and at each stage the state of the element is examined for a nonlinear stress-strain relationship, cracking, yielding of concrete and steel, and the corresponding effects are considered through the appropriate changes in the constitutive relations. The procedure is explained by Krishnamoorthy and Panneerselvam (1978).

3.5 OPTIMAL DESIGN OF REINFORCED CONCRETE FRAMES

Mathematical programming techniques provide powerful tools that can be successfully used to find the best possible design among the many potential designs satisfying the given criteria. With increasing use of computers in design practice, there is growing interest amongst research workers to develop models and programs for optimisation in structural design. A review of these developments has been presented by Krishnamoorthy and Mosi (1979).

For the design of reinforced concrete beams and slabs, Templeman (1972) has used a geometric programming technique. Skelton (1972) developed analytical methods for optimum design of floor systems based on the feasible gradient method.

3.5.1 Linear Programming Model

Cohn and Grierson (1968) have illustrated that the techniques adopted for steel structures are not strictly applicable to reinforced concrete frames. The major difficulties arise from the limit design requirements to satisfy all the three conditions: (a) compatibility and limited ductility; (b) equilibrium; and (c) serviceability conditions under working loads. Unlike standard steel members, the weight and moment capacity of a concrete member varies along the length of the member depending on the detailing of the reinforcement. It may also be noted that the minimum structural weight need not be the only decisive optimality criterion for design. Thus the choice of an objective function is not straightforward as for steel frames.

The LP model proposed by Cohn and Grierson (1968) considers the limit equilibrium and serviceability requirements of the frame as the governing constraints. The total volume of steel reinforcement is expressed as a linear objective function.

An LP model for optimal design of reinforced concrete frames was presented by Munro *et al.* (1972), Krishnamoorthy (1972), and Krishnamoorthy and Munro (1973). The problem is formulated to satisfy (a) compatibility and limited ductility, (b) equilibrium, and (c) serviceability conditions. The optimality criterion used is the total volume of reinforcement and this has been expressed as a linear function after making a suitable approximation to the design of sections. The general limit method explained in Section 3.4.1.1 has been used to express the compatibility and limited ductility criteria. The design bending moment M_d^j at a critical section of the frame under ultimate load is expressed in the following form:

NONLINEAR ANALYSIS AND OPTIMAL DESIGN 97

$$M_d^j = x_j M_u^j \tag{3.35}$$

where x_j is the yield safety parameter and M_u^j is the factored elastic moment. It is assumed that the serviceability requirements are satisfied by specifying a lower bound on the values of x_j, i.e. $x_j \geqslant L_j$.
The LP problem was solved using the standard simplex method.

3.5.2 Nonlinear Programming Formulation

The linear programming models described above do not include the sizes of the concrete sections as design variables, except the area of steel at critical sections. Any rigorous formulation of the optimal design problem becomes a highly nonlinear programming problem with considerable complexity. However, with increasing advances in the numerical algorithms for solution of such problems, it is now possible to formulate the optimal design of reinforced concrete sections and frames as nonlinear problems that satisfy all the design criteria. A brief description of these models is given below.

3.5.2.1 Optimal Design of Beam Sections

Reinforced concrete beam sections can be rectangular, T or L sections. The optimisation problem is formulated as illustrated below, using provisions similar to those specified in the British Code of Practice for concrete structures.

Objective function. The objective function, a minimum value of which is to be found, is the cost of the beam per unit length. For a rectangular section it can be expressed as follows:

$$y = C_c b (d + d') + C_s A_{st} \gamma_s + C_f [b + 2(d + d')] \tag{3.36}$$

Constraints. The following constraints must be satisfied:

1. The section must be adequate to resist the external moment. Thus
$$M_u \geqslant M_d \tag{3.37}$$
Figure 3.13(a) shows the stress and strain distributions across a cross-section. The above condition can be expressed as

$$0.87 f_y A_{st} \left(d - \frac{K_2 \times 0.87 \times f_y A_{st}}{K_1 K_3 f_{ck} b} \right) - M_d \geqslant 0 \tag{3.38}$$

where K_1, K_2 and K_3 are stress block parameters and $\dfrac{f_y}{1.15} = 0.87 f_y$.

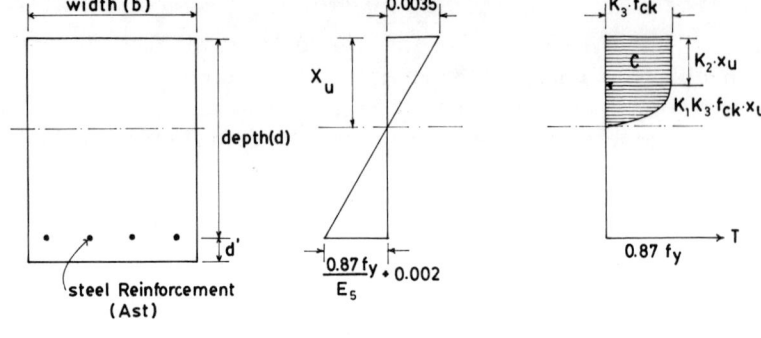

FIG. 3.13a. Singly reinforced beam section.

2. To ensure ductile failure, the beam must be under-reinforced. The limiting percentage of steel, $p_{\lim}$, can be calculated for the balanced condition, and can be expressed as

$$1\cdot 0 - \frac{A_{st}}{bdp_{\lim}} \geqslant 0 \tag{3.39}$$

3. To limit excessive cracking, codes specify a minimum percentage of steel, which is expressed as a constraint:

$$\frac{A_{st}}{bd} \frac{100}{p_{\min}} - 1\cdot 0 \geqslant 0 \tag{3.40}$$

4. To avoid thin sections, a limit is placed on the minimum width:

$$\frac{b}{b_{\min}} - 1\cdot 0 \geqslant 0 \tag{3.41}$$

5. To limit excessive deflections of the beam, the codes require satisfaction of a span to depth ratio, depending on the type of beam. This condition can be written as

$$\frac{S_{di} \times d}{\text{Span}} - 1\cdot 0 \geqslant 0 \tag{3.42}$$

where S_{di} is the span to effective depth ratio as specified in codes of practice and $i = 1, 2, 3$ for cantilever, simply supported and continuous beams respectively, and span is the span length of the beam.

6. To avoid large depths affecting the head room, and also for slenderness considerations, a constraint is imposed by placing a limit on the depth to width ratio:

$$\frac{bd_r}{d} - 1 \cdot 0 \geq 0 \qquad (3.43)$$

Formulation. The optimal design problem to find the variables b, d and A_{st} is defined by minimisation of the objective function (eqn 3.36) subject to constraints (eqns 3.38 to 3.43). It is a nonlinear programming problem (NLP) that can be solved by any of the standard algorithms.

3.5.2.2 Optimal Design of Column Sections

Optimisation of column sections subject to axial load and uniaxial moment is formulated as a nonlinear optimisation problem.

Objective function. The cost per unit length of the column, a minimum value of which is to be found, is given below:

$$y = C_c b (d + d') + C_s A_s \gamma_s + C_f \times 2 [b + (d + d')] \qquad (3.44)$$

Constraints. The following constraints must be satisfied:

1. The section must be capable of carrying the external design forces safely and this constraint can be stated as

$$M_u \geq M_d \qquad (3.45)$$

Due to interaction between axial load and moment, the ultimate moment of resistance M_u cannot be explicitly expressed in terms of the design variables. Figure 3.13(b) shows the stress and strain distributions at the limit state of collapse. It is to be noted that the maximum strain is a function of the depth of neutral axis. From equilibrium considerations it follows that

$$P_u = \left[K_1 K_3 f_{ck} x_u b + \sum_{i=1}^{n_r} A_{si} (f_{si} - f_{ci}) \right] \qquad (3.46)$$

$$M_u = \left[K_1 K_3 f_{ck} x_u b \left(\frac{D}{2} - K_2 x_u \right) + \sum_{i=1}^{n_r} A_{si} (f_{si} - f_{ci}) y_i \right] \qquad (3.47)$$

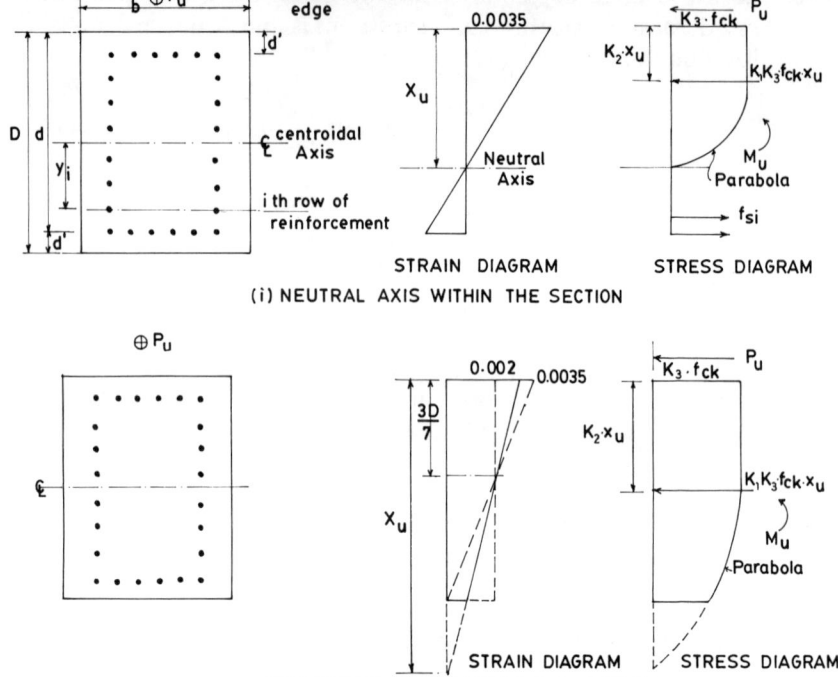

FIG. 3.13b. Column section.

where M_u is the moment about the centroid of the section, y_i is the distance of the ith row of reinforcement from the centroid of the section and n_r is the number of rows of reinforcement.

For a given axial load, the neutral axis depth x_u is fixed iteratively satisfying the force equilibrium equation (3.46), and the ultimate moment capacity is found from eqn (3.47). The constraint of eqn (3.45) is written as

$$\frac{M_u}{M_d} - 1\cdot 0 \geqslant 0 \tag{3.48}$$

2. The lower limit for the allowable percentage of steel not less than 0·8% is satisfied by specifying

$$\frac{A_s}{0\cdot 008\, bD} - 1\cdot 0 \geqslant 0 \tag{3.49}$$

3. To avoid difficulties in placing concrete, the maximum limit for the area of steel is fixed as 4% and this can be stated as

$$\frac{0 \cdot 04\, bD}{A_s} - 1 \cdot 0 \geqslant 0 \quad (3.50)$$

4. A minimum allowable width can be specified:

$$\frac{b}{b_{min}} - 1 \cdot 0 \geqslant 0 \quad (3.51)$$

5. To avoid very deep sections, the depth to width ratio can be limited to an upper bound d_r:

$$\frac{bd_r}{d} - 1 \cdot 0 \geqslant 0 \quad (3.52)$$

where d_r is the maximum depth to width ratio specified.

Formulation. The nonlinear optimisation problem is defined by the objective function (eqn 3.44) and constraints (eqns 3.48 to 3.52).

3.5.2.3 Optimal Design of Frames

A systematic formulation of nonlinear optimisation of reinforced concrete frames has been proposed by Mosi (1980) and the application with inelastic analysis has been investigated by Krishnamoorthy and Mosi (1980). The NLP model is briefly explained below.

In most practical applications, column and beam sections will be kept the same in a storey, or in a certain number of storeys, to minimise construction costs. This aspect is included in the formulation so that the user can specify identical members forming a group and this strategy is termed Design Member Linking Strategy. The design variables are the same for all the members in a group. The design variables chosen for a column group are (a) breadth b (x_1), (b) depth d (x_2) and (c) total area of steel reinforcement A_s (x_3). The design variables chosen for a beam group are (a) breadth $b(x_1)$, (b) depth $d(x_2)$, and (c) area of steel reinforcement A_{st}, placed at the two end sections (x_3, x_5) and midsection (x_4). Thus, a column group has three design variables and a beam group has five design variables.

Objective function. The objective function for a column group is given by

$$y_c = \sum_k \left\{ [C_c x_1 (x_2 + d') + C_3 x_3 \gamma_s + 2 C_f (x_1 + x_2 + d')] L^k \right\} \quad (3.53)$$

where L is the length of the member and k takes appropriate values to include all the members belonging to the group. The objective function for a beam group is given by

$$y_b = \sum_k \left\{ \left[C_c x_1 (x_2 + d') + C_s \gamma_s \left(\frac{x_3}{4} + \frac{x_4}{2} + \frac{x_5}{4} \right) \right. \right.$$
$$\left. \left. + C_f (x_1 + 2x_2 + 2d') \right] L^k \right\} \quad (3.54)$$

Constraints:

Column group. The design constraint is evaluated at the end sections of each member in a column group. This becomes necessary because of axial load and moment interaction. If there are n_1 members in a column group there will be $2n_1$ design constraints to be evaluated. The set of constraints for a column group can be specified following the description given in Section 3.5.2.2.

$$(M_u^k)_j / (M_d^k)_j - 1 \cdot 0 \geqslant 0, \quad j = 1, 2$$

k takes appropriate values to include all members in a group.

$$0 \cdot 04 x_1 (x_2 + d') / x_3 - 1 \cdot 0 \geqslant 0$$

$$\frac{x_3}{0 \cdot 008 x_1 (x_2 + d')} - 1 \cdot 0 \geqslant 0$$

$$\frac{x_1}{(x_1)_{\min}} - 1 \cdot 0 \geqslant 0$$

$$\frac{x_1 d_r}{x_2} - 1 \cdot 0 \geqslant 0 \quad (3.55)$$

In total there are $(2n_1 + 4)$ constraints in a column group.

Beam group. For a beam group, the design constraints are evaluated at the end sections and middle section, for maximum moment to occur at these sections. The set of constraints in which the steel area is involved will have to be evaluated at the three sections. Thus, for a beam group, the constraints are specified as follows:

$$0 \cdot 87 f_y x_i \left(x_2 - \frac{K_2 \times 0 \cdot 87 \times f_y x_i}{K_1 K_3 f_{ck} x_1} \right) / M_d^j - 1 \cdot 0 \geqslant 0$$
$$i = 3, 4, 5 \text{ for } j = 1, 2, 3$$

$$\left.\begin{array}{r}\dfrac{x_1 x_2 p_{\lim}}{100\, x_i} - 1\cdot 0 \;\geqslant\; 0 \\[6pt] \dfrac{x_i}{x_1 x_2}\dfrac{100}{p_{\min}} - 1\cdot 0 \;\geqslant\; 0 \end{array}\right\} \quad i = 3, 4, 5$$

$$\dfrac{x_1}{(x_1)_{\min}} - 1\cdot 0 \;\geqslant\; 0$$

$$\dfrac{x_1 d_r}{x_2} - 1\cdot 0 \;\geqslant\; 0 \tag{3.56}$$

There are 11 constraints in all for a beam group.

Formulation and optimisation. Equations (3.53) to (3.56) should be specified for all beam and column groups in a frame and the problem is to be solved using any NLP technique. In particular, the sub-optimisation procedure using the Sequential Unconstrained Minimisation Technique with Direct Search Method has been reported to be quite adequate for practical solution. Mosi (1980) has also used DFP (Davidon–Fletcher–Powell) algorithm along with SUMT. Elastic analysis has been used for finding the stress resultants under ultimate load. Krishnamoorthy and Mosi (1981) have reported the application of this procedure using inelastic analysis by method of imposed rotations.

3.6 COMPUTER SOFTWARE AND PRACTICAL APPLICATIONS

For practical application of the nonlinear analysis and optimal design methods described in the earlier sections, computer programs must be available to the design engineers. A review of the available software for the analysis and design of reinforced concrete structures was reported by Krishnamoorthy (1982). As it is difficult to describe all the details of the programs available, only a brief description is given below drawing attention to the specific publications on the software.

3.6.1 Nonlinear Analysis of Concrete Framed Structures

The simplified limit method of Baker with the modified procedure by adjustment of rotations has been programmed and the details of the program with practical applications has been reported by Krishnamoorthy and Yu (1972). The program requires input of flexural rigidity EI of members and rotation capacity of the sections along with other geometric details. The program is based on flexibility matrix analysis and is suitable for regular rectangular frameworks.

The method of imposed rotations due to Macchi has been formulated as a linear complimentary program by Maier *et al.* (1972), and improved computer implementation with flow chart has been presented by Krishnamoorthy and Mosi (1980). The program is based on stiffness analysis and shows promise for wider application. The program requires input of idealised moment–rotation characteristics of critical sections besides the usual geometric and other data.

The nonlinear finite element analysis application is still in the research stage. Though successful applications have been reported by Krishnamoorthy and Panneerselvam (1978), it is costly to run such programs. However, simplified finite element models may be found economical for practical applications.

3.6.2 Optimal Design Software

One of the difficulties in the development of design software is that it becomes Code dependent and cannot be used in other countries without adjustments. Hence, based on the accepted formulations, programs need to be developed to suit the requirements of Code specifications in a particular country. The nonlinear optimisation models described in Section 3.5 are mostly general and can be used with proper values of the stress block parameters specified in a Code of Practice.

The SUMT algorithm with direct search method or DFP method have been found to be quite adequate to develop programs for optimal design of reinforced concrete beam and column sections. These programs can be used to draw design charts for optimal design, for given values of cost of concrete, steel and formwork, for various values of moment, axial load and moment.

The design member linking strategy adopted by Mosi (1980) is quite effective in reducing the number of design variables in frame optimisation. Computer programs have been developed using elastic analysis and inelastic analysis based on the method of imposed rotations. Figure 3.14 shows examples of frames reported by Mosi (1980) and Krishnamoorthy and Mosi (1981). The SUMT algorithm with direct search method has been found to be simple and adequate for the problems investigated.

NONLINEAR ANALYSIS AND OPTIMAL DESIGN 105

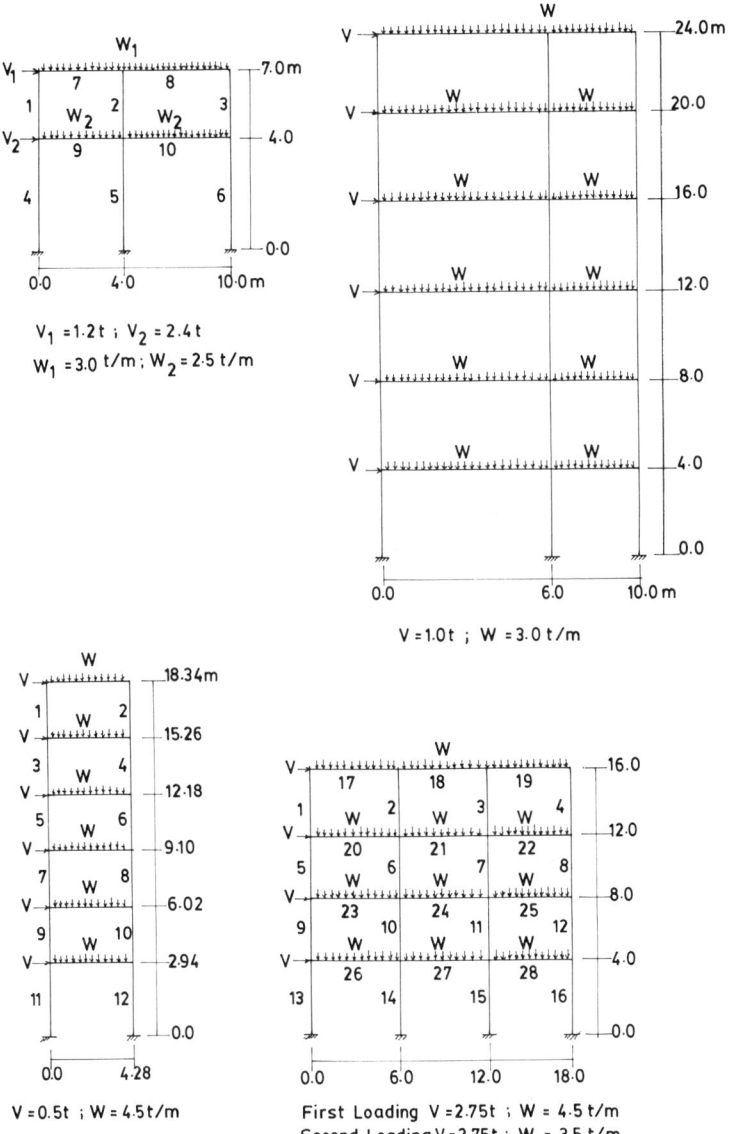

FIG. 3.14. Typical frames.

3.7 CONCLUSIONS AND FUTURE RESEARCH

In this chapter the important methods of nonlinear analysis and formulations for optimal design of reinforced concrete frames are described. In order to put these methods into wider use, there are several areas which require further research:

1. (a) Nonlinear analysis is required for rational assessment of forces in reinforced concrete frames under limit state of collapse and serviceability conditions.
   (b) The simplified limit method with the modified procedure can be adopted for analysis of simple and regular frames.
   (c) The method of imposed rotations formulated as a linear complementarity problem provides a computationally efficient procedure for analysis of frames for limit state of collapse and also for service load conditions.
2. (a) The application of inelastic analysis procedures requires more realistic data on flexural rigidity of members. Experimental investigations oriented to provide simplified expressions for computing EI values for use in the above methods are required.
   (b) Though a large number of experimental investigations have been carried out, further work is necessary to propose suitable mathematical models for trilinear moment–rotation relations for various axial load and moment interactions, and percentage of steel. Specifically these relations must fit into the computer implementation of the method of imposed rotations.
   (c) In the case of frames subject to large sidesway, the second-order P-δ effects and instability require particular attention, especially under ultimate load conditions when a number of critical sections develop inelastic rotations. Nahhas and Yu (1972) have suggested a simplified procedure, based on analytical study, to magnify the lateral loads to safeguard against instability.
   (d) Detailed large scale experimental studies need to be planned and conducted on braced and unbraced frames typical of multistorey buildings subject to loading conditions specified in various Codes of Practice. The results of these investigations would provide the data necessary to develop and validate the mathematical models and analysis procedures for limit state design of frames.

3. Nonlinear finite element analysis provides a powerful and rigorous tool for analysis of reinforced concrete structures. It can be expected that the inclusion of the effect of reinforcement as an integral part of the element and the use of actual stress–strain relations and cracking of concrete will provide the necessary analytical tool, which is lacking so far, to compute the stress resultants at a particular stage of loading.

   For application to large frames, simpler models are needed, such as the layered finite elements proposed by Franklin (1970) but refined to include effects due to shear. Sub-structuring techniques for nonlinear analysis may prove to be effective and require investigation for application to multistorey framed structures.

4. In many practical applications, the frames form an integrated three-dimensional structural system consisting of beams, columns, floor slabs and shear walls. Even the three-dimensional finite element analysis assuming linear behaviour with a number of simplifications (e.g. the floor systems to behave like rigid diaphragms, etc.) is quite tedious and costly. The order of magnitude of the complexity involved increases considerably for nonlinear analysis of such systems. Though these developments are not of immediate practical use, they provide vast scope and potential for future research into finite element models and programs for rigorous analysis of the integrated behaviour of these systems.

5. (a) The linear programming model for frame optimisation provides a simpler algorithm but it has the limitation that it can be used to optimise only the total volume of reinforcement, after specifying the values of the cross-sectional dimensions of the members.

   (b) Nonlinear programming formulations show promise for wider use in optimal design of reinforced concrete beam and column sections and also frames. However, to improve the computational efficiency and for applications to large frames, further research is needed.

   (i) Nonlinear programming formulations should also include deflection constraints to satisfy serviceability constraints and $P\text{-}\delta$ effects in the case of large frames.

   (ii) Numerical investigation is required on several solution algorithms to identify computationally efficient algorithms for practical applications.

(iii) For optimisation of large frames, a substructuring technique seems to be necessary and this has to be investigated in the context of nonlinear optimisation with nonlinear analysis procedure.

(c) Developing an optimal design software package that can be used in all countries is difficult. A suitable package has to be designed in such a way that subsystems for analysis and optimisation problem solver routines form a common core, with user oriented interfaces developed to take care of design requirements in various Codes of Practice.

REFERENCES

BAKER, A. L. L. (1956) *Ultimate Load Theory Applied to the Design of Reinforced and Prestressed Concrete Frames,* Concrete Publications, London.

BAKER, A. L. L. (1970) *Limit State Design of Reinforced Concrete,* Cement and Concrete Association, London.

BAKER, A. L. L. and AMARAKONE, A. M. N. (1964) Inelastic hyperstatic frames analysis. *Proc. Int. Symposium on Flexural Mechanics of Reinforced Concrete, Miami, Florida,* ACI Special Publication SP-12, pp. 85–142.

CEB-FIP (1970) *International Recommendations for the Design and Construction of Concrete Members,* Comité Européen du Béton-Fédération Internationale de la Précontrainte, Paris. English Translation from Cement and Concrete Association, London.

CHEN, W. F. (1982) *Plasticity in Reinforced Concrete,* McGraw-Hill Book Company, NY.

COHN, M. Z. (1964) Why nonlinear analysis and design? *Proc. Int. Symposium on Flexural Mechanics of Reinforced Concrete, Miami, Florida,* ACI Special Publication SP-12, pp. 591–4.

COHN, M. Z. (Ed.) (1972) *Inelasticity and Nonlinearity in Structural Concrete,* Proc. Symposium, University of Waterloo, Canada, Study No. 8, Solid Mechanics Division.

COHN M. Z. and GRIERSON, D. E. (1968) Optimal design of reinforced concrete beams and frames. *Final Publication, 8th IABSE Congress,* pp. 215–26.

COLVILLE, J. and ABBASI, J. (1974) Plane stress reinforced concrete finite elements. *J. of the Structural Division, ASCE,* **100** (ST5), 1067–83.

DE DONATO, O. and MAIER, G. (1972) Mathematical programming methods for the inelastic analysis of reinforced concrete frames allowing for limited rotation capacity. *Int. J. for Num. Meth. in Eng.,* **4**, 307–29.

FORD, J. S., CHANG, D. C. and BREEN, J. E. (1981) Experimental and analytical modelling of unbraced multipanel concrete frames. *ACI Journal,* **78**, 21–35.

FRANKLIN, H. A. (1970) Nonlinear Analysis of Reinforced Concrete Frames and Panels. PhD Dissertation, Division of Structural Engineering and Structural Mechanics, University of California, Berkeley.

HOGNESTAD, E., HANSON, N. W. and McHENRY, D. (1955) Concrete stress distribution in ultimate strength design. *ACI Journal,* **52**, 455-79.
KANEKO, I. (1977) A mathematical programming method for the inelastic analysis of reinforced concrete frames. *Int. J. for Num. Meth. in Eng.,* **11**, 1137-54.
KRISHNAMOORTHY, C. S. (1972) Optimal Limit Design of Reinforced Concrete Frames. PhD Thesis, Imperial College of Science and Technology, London.
KRISHNAMOORTHY, C. S. (1982) *Analysis and Design of Reinforced Concrete Structures,* Structural Mechanics Software Series, Edited by PERRONE, N. and PILKEY, W., vol. iv, University Press of Virginia, Charlottesville, USA, pp. 239-46.
KRISHNAMOORTHY, C. S. and MOSI, D. R. (1979) A survey on optimal design of civil engineering structural systems. *Engineering Optimization,* **4** (2), 73-88.
KRISHNAMOORTHY, C. S. and MOSI, D. R. (1980) CONFAP — A computer program for inelastic analysis of reinforced concrete framed structures. *Computers and Structures,* **12**, 677-87.
KRISHNAMOORTHY, C. S. and MOSI, D. R. (1981) Optimal design of reinforced concrete frames based on inelastic analysis. *Engineering Optimization,* **5** (3), 151-67.
KRISHNAMOORTHY, C. S. and MUNRO, J. (1973) Linear program for optimal design of reinforced concrete frames. *IABSE Publications,* **33**, 119-41.
KRISHNAMOORTHY, C. S. and PANNEERSELVAM, A. (1977) A finite element model for nonlinear analysis of reinforced concrete framed structures. *The Structural Engineer (London),* **55** (8), 331-8.
KRISHNAMOORTHY, C. S. and PANNEERSELVAM, A. (1978) FEPACS1 — A finite element program for nonlinear analysis of reinforced concrete framed structures. *Computers and Structures,* **9**, 451-61.
KRISHNAMOORTHY, C. S. and YU, C. W. (1972) Simplified computer approach to the ultimate load analysis and design of reinforced concrete frames. *ACI Journal,* **69**, 690-8.
KUPFER, H. B., HILSDORF, H. K. and RUSCH, H. (1969) Behaviour of concrete under biaxial stresses. *ACI Journal,* **66**, 656-66.
MACCHI, G. (1966) Methode des rotations imposees. *Structures Hyperstatiques,* **53**, CEB (Project d'annexe aux recommendations pratiques).
MACCHI, G. (1969) Limit states design of statically indeterminate structures composed of linear members. *Studie Rendiconti del Corso de Perfezionamento di Construzioni in Cemento armato,* **6**, 151-91.
MACCHI, G. and SIVIERO, E. (1974) Deformability of prismatic R.C. members with rectangular cross-section under combined bending and axial load. *CEB, Bulletin D' Information,* No. 101, pp. 191-202.
MAIER, G., DE DONATO, O. and CORRADI, L. (1972) Inelastic analysis of reinforced concrete frames by quadratic programming. *Inelasticity and Nonlinearity in Structural Concrete,* Edited by COHN, M. Z., Proc. Symposium, University of Waterloo, Canada, Study No. 8, pp. 265-88.
MOSI, D. R. (1980) Nonlinear Optimization of Reinforced Concrete Frames. PhD Thesis, Dept of Civil Engineering, Indian Institute of Technology, Madras, India.

MUNRO, J. (1965) The elastic and limit analysis of planar skeletal structures. *Civil Engineering and Public Works Review (London)*, **60** (706), 671–7.
MUNRO, J., KRISHNAMOORTHY, C. S. and YU, C. W. (1972) Optimal design of reinforced concrete frames. *The Structural Engineer (London)*, **50** (7), 259–64.
NAHHAS, U. and YU, C. W. (1972) The elastic–plastic design of reinforced concrete sway frames against instability. *Proc. Inst. of Engineers (London)*, **53** (2), 41–56.
NGO, D. and SCORDELIS, A. C. (1967) Finite element analysis of reinforced concrete beams. *ACI Journal*, **64**, 152–63.
PANNEERSELVAM, A. (1977) Nonlinear Finite Element Analysis of Reinforced Concrete Framed Structures. PhD Thesis, Dept of Civil Engineering, Indian Institute of Technology, Madras, India.
RÜSCH, H. (1960) Researches towards a general flexural theory for structural concrete. *ACI Journal*, **57**, 1–28.
SARGIN, M. (1971) *Stress–Strain Relationships for Concrete and the Analysis of Structural Concrete Sections,* Study No. 4, Solid Mechanics Division, University of Waterloo, Canada, p. 167.
SCORDELIS, A. C. (1972) Finite element analysis of reinforced concrete structures. *Proc. Speciality Conference on the Finite Element Methods in Civil Engineering,* Montreal, Canada, pp. 71–113.
SKELTON, R. (1972) Analytical Methods for the Optimum Design of Reinforced Concrete Slab and Beam Structures. PhD Thesis, Queen's University, Belfast.
TEMPLEMAN, A. B. (1972) Geometric programming with examples of the optimum design of floor and roof systems. *Proc. Int. Symposium on Computer-Aided Structural Design,* University of Warwick, England.
WINTER, G. (1964) Whither inelastic concrete design. *Proc. Int. Symposium on Flexural Mechanics of Reinforced Concrete, Miami, Florida,* ACI Special Publication SP-12, pp. 581–9.

Chapter 4

REINFORCED CONCRETE COLUMNS IN BIAXIAL BENDING

Issam E. Harik and Hans Gesund
Department of Civil Engineering, University of Kentucky, Lexington, Kentucky, USA

SUMMARY

Methods are presented for the analysis and design of reinforced concrete columns subjected to combined axial load and biaxial bending. They range from the most burdensome, but presumably most accurate, to relatively simple procedures which have given acceptable results. Determination of the stiffness matrices of such members is also an important part of the analysis, since they are needed to find the forces for which the members must be designed and analyzed. Both 'short' and 'slender' columns are considered, and the frames of which they are a part may be braced or unbraced.

NOTATION

b	Width of concrete cross section
b_w	Width of stem of beam cross section
C_c	Resultant compressive force in the concrete
C_m	Factor for moment magnification
C_{si}	Compressive force in bar i
E_c, E_s	Modulus of elasticity for concrete and steel
e_{0x}, e_{0y}	Uniaxial eccentricities when M_{ny} and M_{nx} are zero respectively
h	Thickness of concrete cross section
I_g	Moment of inertia of gross concrete section neglecting reinforcement

I_{se}	Moment of inertia of reinforcement
k	Effective length factor (could be determined from alignment charts or from equations in Figs 4.4 and 4.5)
l_u	Unbraced length
M_{0x}, M_{0y}	Uniaxial moment strengths at failure load P_n when M_{ny} and M_{nx} are zero respectively
M_{1b}, M_{2b}	End moments resulting from no appreciable sidesway (i.e. gravity load moments). $M_{1b} < M_{2b}$ and M_{1b}/M_{2b} is positive for single curvature
M_{1s}, M_{2s}	Sidesway related end moments. $M_{1s} < M_{2s}$
M_c	Magnified factored (ultimate) moment
M_n, M_u	Nominal moment strength and factored (ultimate) moment
P_c	Euler buckling load; ΣP_c (summation for all columns in a storey)
P_n, P_u	Nominal axial load strength and factored axial force; ΣP_u (summation for all columns in a storey)
P_0	Load carrying capacity under pure axial compression
P_{0x}, P_{0y}	Load carrying capacities under compression with uniaxial eccentricities e_{0x} and e_{0y} respectively
r	$\sqrt{I_g/A_g}$; radius of gyration
T_{sj}	Tensile force in bar j
x, y	Subscripts denoting bending about x and y axes
β_d	$\left[\dfrac{\text{maximum factored dead load moment}}{\text{maximum factored total load moment}}\right]$; absolute value of portion of factored load moment that is sustained
δ_b	Magnification factor for a braced frame compression member
δ_s	Magnification factor for an unbraced frame compression member to account for $P\text{-}\Delta$ effect
ρ_g	Ratio between area of longitudinal reinforcement and gross area of column cross section
ϕ	ACI code strength reduction factor. $\phi = 0.70$ for tied compression members, $\phi = 0.75$ for spirally reinforced members
ψ	$(\Sigma EI/l$ for compression members$)/(\Sigma EI/l$ for flexural members), end restraint factor; ψ_A (for end A of member AB); ψ_B (for end B of member AB)
ψ_a	Average of ψ values at ends of column
ψ_{min}	Minimum of ψ values at ends of column

4.1 INTRODUCTION

The design or analysis of a reinforced concrete column must be undertaken in conjunction with that of the entire structure. In general the same simplifying assumptions must therefore be used. These may range from using the same EI for all members in a frame, ignoring size and reinforcing variations, to only smearing composite nominal tangent EIs over short segments of each member, while taking into account variations in cracking and load levels, as well as the effects of shrinkage, creep, reinforcement distribution and the nonlinear short time stress–strain responses of the materials.

Using any desired assumptions, it is necessary to obtain a member stiffness matrix in order to find the axial load and moments and shears which the member will have to resist prior to collapse at ultimate load, and with deformations and crack widths less than their respective permissible values at service loads. Member cross sections designed to accomplish this can then be used to obtain a new stiffness matrix for use in a new cycle of analysis and design.

The shear strength, shown to be a factor of utmost importance in earthquake resistance, is traditionally dealt with separately. The fact that there is biaxial rather than uniaxial bending in the column need not change the shear analysis. The member must be analyzed for the resultant shear force.

Given the previously stated requirements, it is obvious that the mere design of the column cross section for strength under some combination of 'ultimate' loads is quite insufficient. However, the strength and behavior of the cross section represent anchor points in the analysis and design of reinforced concrete columns. In this chapter a variety of methods for the design and analysis of slender and nonslender reinforced concrete columns will be presented and examined, some simple and inaccurate, others more complex and, hopefully, more accurate. Given the nature of the problem, a combination of simplicity and accuracy is probably impossible to achieve, though it appears that satisfactory performance of columns can be obtained quite economically.

The design and analysis of reinforced concrete columns under combined axial load and uniaxial bending is discussed in many textbooks on reinforced concrete structures (Wang and Salmon, 1985). Biaxial bending introduces complications in inverse ratio to the number of simplifying assumptions to be made in the structural analysis. If the simplest method is used, in which fully elastic, uncracked, short column behavior is assumed, neither the presence of axial load nor that of bending in the orthogonal direction will affect the stiffness of the member in one direction, and cross section stresses may be obtained from uncoupled bidirectional frame analyses and superposed. Many columns designed in this way are serving successfully all over the world.

At the other end of the spectrum of complexities, the length of the column is divided into short segments and a grid is superimposed on the cross section of each. The time and stress dependent responses of both the concrete and the individual reinforcing bars are tracked in a historical loading sequence, and axial load dependent, biaxial moment-curvature relationships of each segment are calculated. Numerical integration over the member length is then used to obtain a member stiffness matrix for just that history and set of loading conditions. Enormous computer resources are required to iterate through an entire structural frame in this manner and no real building columns have yet been designed using it. Very small frames have, however, been analyzed by Gesund (1967) and Shah and Gesund (1972).

Between the two extremes, a number of useful methods of design, analysis, redesign, reanalysis, etc., have been developed. Several have proven to yield safe and economical designs in practice.

Almost all Codes of Practice require design for 'ultimate' loads, but recommend that an elastic method of frame analysis be used. This defies logic, since the structure cross sections will be anything but elastic at the loads for which the analysis is being carried out. They will, in fact, have been designed for large yielding of the reinforcement and incipient crushing of the compression concrete, i.e. considerable hinge rotations. Nevertheless, for frame members not affected by slenderness and/or sidesway deformations, the method has been found to be workable though, lacking controls on hinge rotations, not always conservative, and it has permitted the economical design of a large number of serviceable structures. It is doubtful that it should be used with the same confidence when story drifts and member or frame stability become considerable problems.

4.2 NONSLENDER COLUMNS

4.2.1 Analysis

Columns in which slenderness effects need not be considered are sometimes referred to as short columns, column segments, zero length

REINFORCED CONCRETE COLUMNS IN BIAXIAL BENDING

columns, or columns with sufficient lateral bracing. The strains, stresses and resultant forces acting on a biaxially loaded column cross section, at some stress level, are shown in Fig. 4.1 for a given neutral axis location and orientation. x_c and y_c are the coordinates of the centroid of the concrete compression stress block, while x_{si} and y_{si} are the coordinates of the centroid of bar i. When the strains due to shrinkage in the concrete are considered, the zero stress axis in the concrete will not coincide with the zero strain axis. This is shown in Fig. 4.1. Using notation[†] of Fig. 4.1, the equilibrium equations can be written as follows:

$$P_n = C_c + \Sigma C_{si} - \Sigma T_{sj} \quad (4.1)$$

$$M_{nx} = P_n e_{ny} = C_c y_c + \Sigma C_{si} y_{si} - \Sigma T_{sj} y_{sj} \quad (4.2)$$

$$M_{ny} = P_n e_{nx} = C_c x_c + \Sigma C_{si} x_{si} - \Sigma T_{sj} x_{sj} \quad (4.3)$$

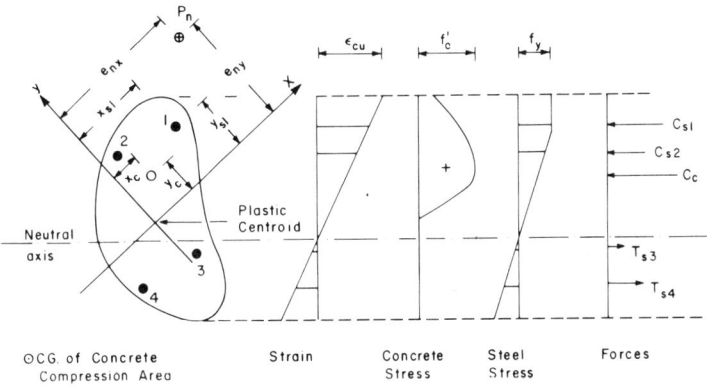

FIG. 4.1. Strains, stresses, and resultant forces on a biaxially loaded cross section.

P_n is the axial force in the column, acting at nominal eccentricities e_{ny} and e_{nx} with respect to the x and y axes. The resultant force in the concrete, C_c, depends on the shape and dimensions of the compression area and on the shape of the concrete stress–strain relationship. C_{si} and T_{sj} are the compressive and tensile forces in bars i and j respectively.

The 'exact' analytical method presented in eqns (4.1)–(4.3) is applicable to reinforced concrete sections of any shape and reinforcement pattern. Any stress–strain relationships may be used for the steel

[†]See also the Notation section, pp. 111–12

and concrete. Other analytical methods have been presented by Au (1958), Czerniak (1962) and Pinto de Magalhaes (1979). Patel and Gesund (1965) avoided some of the mathematical difficulties associated with the analytical methods by imposing a rectangular grid on the cross section. Each rectangular element in the compression zone was assumed to be subjected to uniform compressive stress. A similar procedure was later used by Chen and Shoraka (1975).

The purely analytical methods as well as the grid methods can give direct solutions for moments and axial loads corresponding to a particular strain distribution. They also give the relationships between loads and deformations at any strain or loading level, and therefore the axial load and bending resistances of a column cross section about a particular skewed axis. However, they require a trial and adjustment procedure to obtain the location and inclination of the neutral axis for a given eccentric load.

If they are to be applied to design, these procedures become quite involved, and they are probably not suitable for every day office use. Approximate methods are therefore normally employed for the design process, and frequently also for analysis. These require separation of the strength calculations from the determination of the load–deformation response, i.e. the 'stiffness'.

4.2.2 Design

Several approximate methods have been developed to predict the strength of column cross sections subjected to combined axial load and biaxial bending. The most commonly used is based on the concept of failure surfaces introduced by Bresler (1960) and Pannell (1960, 1963). Three types of failure surfaces (Fig. 4.2) can be produced by plotting the failure load P_n as a function of its eccentricities e_{nx} and e_{ny} or its associated bending moments M_{nx} and M_{ny}. Bresler presented interaction equations for the reciprocal load method (eqn 4.4) based on the planar approximation to the failure surface $S_2(1/P_n, e_{nx}, e_{ny})$, which is the inverse of surface S_1, and for the more accurate load contour method (eqn 4.5) based on the failure surface $S_3(P_n, M_{nx}, M_{ny})$.

$$\frac{1}{P_n} = \frac{1}{P_{0x}} + \frac{1}{P_{0y}} - \frac{1}{P_0} \qquad (4.4)$$

$$\left(\frac{M_{nx}}{M_{0x}}\right)^{a_1} + \left(\frac{M_{ny}}{M_{0y}}\right)^{a_2} \leqslant 1\cdot 0 \qquad (4.5)$$

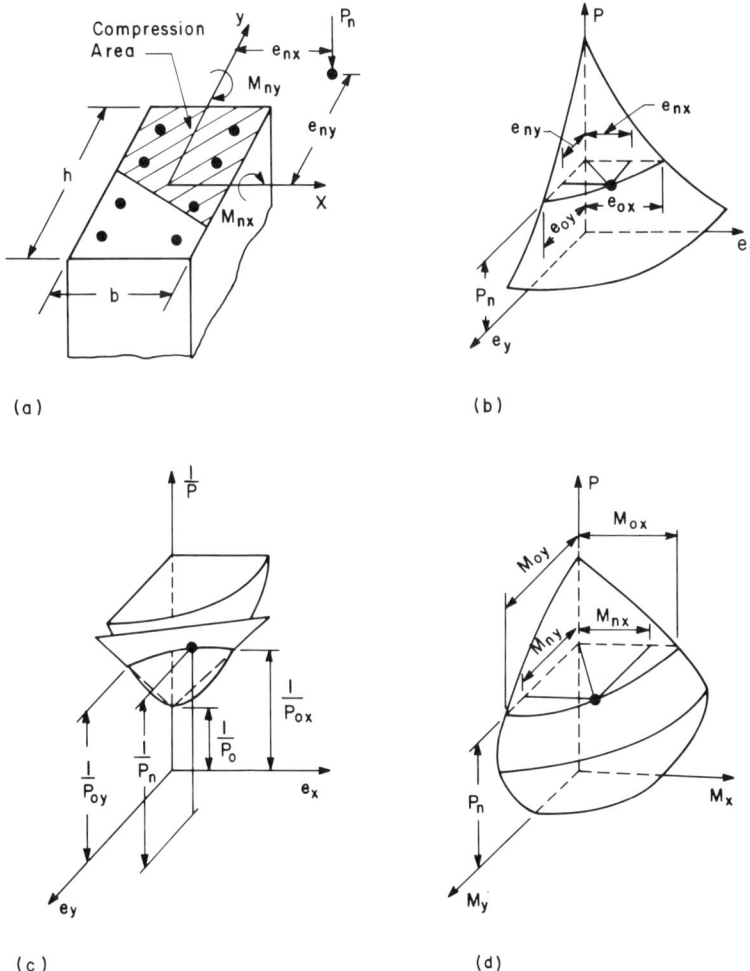

FIG. 4.2. Failure surfaces for a column subjected to biaxial bending (a). (b) Failure surface S_1 (P, e_x, e_y); (c) failure surface S_2 ($1/P$, e_x, e_y); (d) failure surface S_3 (P, M_x, M_y); (after Bresler, 1960).

In these expressions P_{0x} and P_{0y} represent the axial load carrying capacities with uniaxial eccentricities e_{0x} and e_{0y}. P_0 is the load carrying capacity under pure axial compression. M_{0x} and M_{0y} are the uniaxial moment strengths at failure load P_n when applied moments M_{ny} and

M_{nx} are assumed to be zero respectively. The exponents α_1 and α_2 depend on the column dimensions, amount and distribution of reinforcement, stress–strain relationships for steel and concrete, amount of concrete cover, and arrangement and size of lateral ties. Based on tests carried out on five rectangular columns, Bresler suggested that the strength could be closely approximated by assuming $\alpha_1 = \alpha_2$ in eqn (4.5). Parme et al. (1966) suggested the use of $\alpha_1 = \alpha_2 = \log 0.5/\log \beta$, where $M_{nx} = \beta M_{0x}$ and $M_{ny} = \beta M_{0y}$. Gouwens (1975) modified eqn (4.5) and presented analytical expressions for the coefficient β. Figure 4.3 outlines the procedure proposed by Gouwens for the strength design of short columns in biaxial bending. Other strength design or analysis procedures, based on approximations of the failure surfaces, have been presented by Aas-Jakobsen (1964), Furlong (1961, 1979), Meek (1963), Ramamurthy (1966), Shanmughasundaram (1977), and Weber (1966). Marin (1979) and Ramamurthy and Hafeez Khan (1983) presented a design procedure for L-shaped column cross sections and Furlong (1979) analyzed cross sections with rounded ends.

4.2.2.1 Example
Gouwens (1975) with dimensions converted to the SI system. Design a column cross section for the following:

$P_n = 890$ kN, $\quad M_{nx} = 102$ kNm, $\quad M_{ny} = 475$ kNm, $\quad f'_c = 41.4$ MPa, $f_y = 414$ MPa

Solution: (See flowchart of Fig. 4.3)
1. Choose a 508 × 508 mm cross section.
2. Try twelve No. 10 bars (bar area = 820 mm^2) placed evenly on all four sides with

$$\rho = \frac{(12)(820)}{(508)(508)} = 0.0381$$

$M_{0x} = M_{0y} = 559$ kNm

3. $q = 0.0381 \dfrac{414}{41.4} = 0.381 < 0.5$

4. $\beta_1 = 0.545 + 0.35(0.5 - 0.381)^2 - 0 = 0.550$

5. $n = \dfrac{890\,000}{41.4(508)(508)} = 0.0833 < 0.25$

REINFORCED CONCRETE COLUMNS IN BIAXIAL BENDING 119

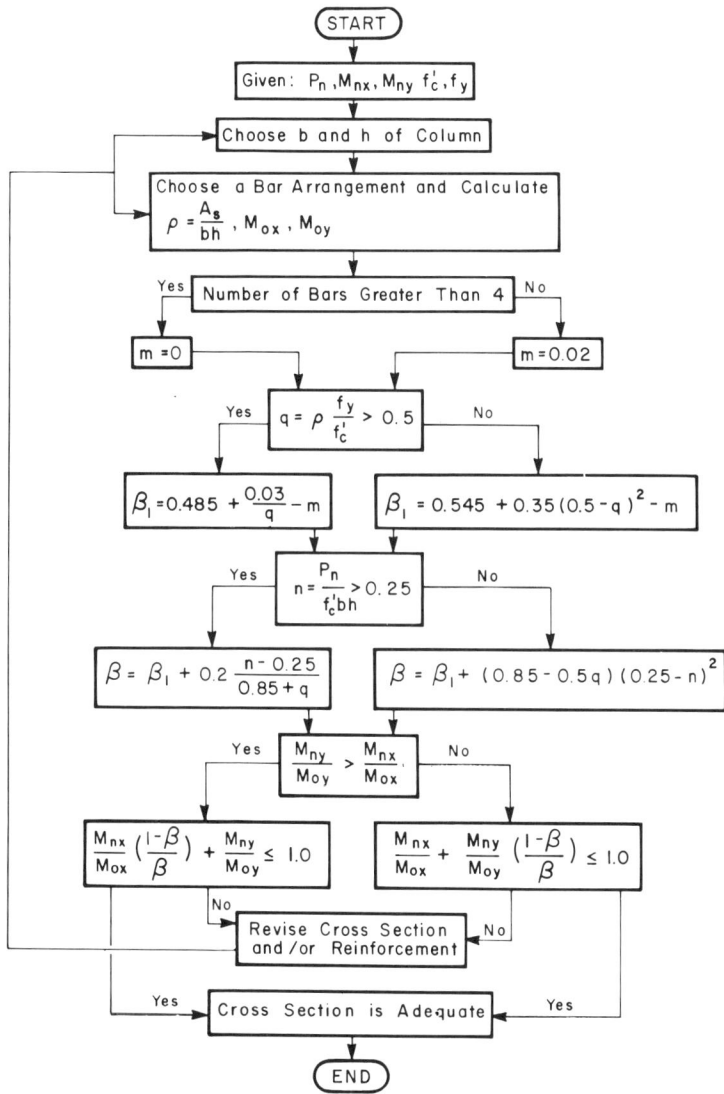

FIG. 4.3. Flowchart for design of column cross sections for combined axial load and biaxial bending.

6. $\beta = 0.550 + [0.85 - (0.5)(0.381)](0.25 - 0.0833)^2 = 0.568$

7. $\dfrac{M_{nx}}{M_{0x}} = \dfrac{102}{559} = 0.182 < \dfrac{M_{ny}}{M_{0y}} = \dfrac{475}{559} = 0.850$

8. $0.182 \left(\dfrac{1 - 0.568}{0.568} \right) + 0.850 = 0.99 < 1.0$

Therefore the design is adequate.

4.2.3 Stiffness

The stiffness of short columns can be obtained either very approximately with little effort, or with the same methods and assumptions used for slender columns (see Section 4.3).

In determining very approximate stiffnesses of the members in a reinforced concrete frame, it is desirable to use consistent assumptions. Thus if beam stiffnesses are to be based on gross moment of inertia, so should column stiffnesses. For braced frames in which column slendernesses will not cause problems, the Commentary to ACI 318-83 suggests use of one half of the gross EI of the beam stems, and the full gross EI of the columns. When using such a crude approximation, which also omits axial load effects and stress levels, the refinement of considering the effect of bending in the orthogonal plane on the EI or stiffness of a column in a given plane, appears difficult to justify. Nevertheless, since Gesund and Vandevelde (1973) found that orthogonal moments would cause considerable reductions in stiffness at, or only a little above, service load levels, it would appear sensible to arbitrarily reduce the column EI when appreciable orthogonal moments are present. The percentage reduction could be the same as the reduction in uniaxial moment capacity due to the presence of the orthogonal moment.

4.3 SLENDER COLUMNS

When slenderness is to be considered in columns subjected to biaxial bending, or when the stiffness is to be determined more accurately for other reasons, it becomes necessary to consider the influence of axial load and orthogonal bending moments on the curvatures and stiffness of the member in a given plane. Gesund and Vandevelde (1973) showed that such orthogonal moments greatly increase curvatures and hence reduce the stiffness of even short columns as loadings approach the ultimate values for which the members must be designed and for which the structure must therefore be analyzed. The effects of creep and shrinkage on the curvatures were also seen to be large. If the distribution of moments and forces in the structure, and therefore the moments and forces acting on a given column, are to be determined with any accuracy, these factors should be included in the member stiffness calculations. This cannot be done cheaply, and may well require a step-by-step historical analysis of a frame and its members (Shah and Gesund, 1972).

4.3.1 Analysis

A comprehensive procedure, embodying the above requirements, has been developed. It provides the column stiffness matrix under all conditions of loading and deformation in the material and the members (Shah and Gesund, 1972; Gesund and Vandevelde, 1973). The procedure is outlined in the following steps:

1. Given a column A–B with known cross section and end loadings, P, M_{Ax}, M_{Ay}, M_{Bx} and M_{By}.
2. Divide the column into short segments. Twenty segments are advisable if the column may be bent into an S shape, otherwise 10 should be sufficient.
3. For each segment carry out the following:
   (a) Impose a rectangular grid on the cross section.
   (b) Assume a neutral axis location and orientation.
   (c) Choose a value for compression strain, including shrinkage and creep, at the center of the grid furthest from the neutral axis.
   (d) Determine the curvatures Γ_x and Γ_y about the x and y axes.
   (e) Assuming linear strain distribution, determine the strains in all bars and at the center of all grid rectangles in the compression area. If a column shortening term is to be included in the stiffness matrix, total vertical strain is also calculated at the center of the column.
   (f) Using actual stress–strain relations of the steel and concrete, including nonlinear and shrinkage and creep effects, obtain stresses and forces in each bar and rectangle.
   (g) Summing forces and moments, determine the axial load P and the bending moments M_x and M_y that the cross section is resisting for the assumed strain distribution.

(h) Choose a new value for the compression strain, and repeat the procedure from step (d) above. After the last desired strain has been used, a new location and orientation of the neutral axis are chosen and the procedure is repeated from step (c). This must be done as often as required for reasonable coverage of the spectrum of applied loads, biaxial moments, and associated curvatures and axial shortening.

(i) The results of the calculations are arranged in tables of $\{P\} \equiv \{M_x\} \equiv \{M_y\} \equiv \{\Gamma_x\} \equiv \{\Gamma_y\}$, for all segments. These tables can be used to generate the equations for five dimensional surfaces representing the cross section responses.

4. Assume an initial biaxially deflected shape for the column. (Frequently, zero displacement at all points will serve well.)

5. Determine biaxial moments in each segment, including those due to the interaction of the axial load and the biaxial displacements.

6. Obtain biaxial curvatures for each segment from the tables or equations generated in step 3(i).

7. Numerical integration of the segment curvatures, or substitution into a finite difference form of the differential equations of the elastic curves, give new displacements of the segments in the biaxial directions. If differences between these displacements and the ones assumed in step 4 are greater than some acceptably small value, use the new displacements and repeat steps 5 to 7 until desired convergence is achieved.

8. Obtain the end rotations θ_{Ax}, θ_{Ay}, θ_{Bx} and θ_{By} from the displacements adjacent to the column ends.

9. Changing the end loadings to P, $(M_{Ax} + \Delta M_{Ax})$, M_{Ay}, M_{Bx}, and M_{By}, in which ΔM_{Ax} is a small increase in M_{Ax}, one can obtain changes in end rotations, $\Delta\theta_{Ax,Ax}$, $\Delta\theta_{Ax,Ay}$, $\Delta\theta_{Ax,Bx}$ and $\Delta\theta_{Ax,By}$. Dividing each of these changes in end rotation by ΔM_{Ax} gives the first column of the column flexibility matrix. Next, increasing end moment M_{Ay} by ΔM_{Ay}, a small increase in M_{Ay}, while holding all other loadings at their original values, one can obtain changes in end rotation $\Delta\theta_{Ay,Ax}$, $\Delta\theta_{Ay,Ay}$, $\Delta\theta_{Ay,Bx}$ and $\Delta\theta_{Ay,By}$. When divided by ΔM_{Ay}, these become the second column of the flexibility matrix. The other two columns are obtained similarly by operating on the moments at B. The column shortening term can be obtained by varying the axial load independently of the moment.

10. Inversion of the flexibility matrix then gives the stiffness matrix.

The use of tables in step 6 will usually require some interpolation, which may cause difficulties. These stem from the fact that the table contains five columns, which complicates the choice of the correct data set, as well as the interpolation. One solution is to require that the data sets (rows) in the table be sufficiently close together so that a search with any entry number or numbers will locate a unique data point within an acceptable margin of error. To avoid the derivation and use of lengthy tables, Shah and Gesund (1972) devised a planar interpolation algorithm. They incorporated the above procedure, as well as a similar but simpler one for beams, as subroutines into an overall three dimensional structural analysis program. In hindsight, it appears that use of an algorithm and program which would fit an equation to the five dimensional table would have been more efficient than interpolation. Al-Noury and Chen (1982) presented a finite segment method in which, instead of setting up a stiffness matrix for the entire column, stiffness matrices were developed for the individual column segments which could then be treated as finite elements in an overall structural analysis.

4.3.2 Design

The above procedures require that the columns be completely designed and detailed before they can be accurately analyzed. The material properties, and perhaps a loading history, must also be known or assumed. Further, a considerable computational effort is required. They are therefore not well suited for design office use, except in very special circumstances.

In order to simplify the analysis and design of slender columns, the current reinforced concrete building and highway bridge codes approximate the effect of slenderness by using the same moment magnifier approach which has been traditionally used in the steel design codes. In this approach, the sum of the primary and secondary (P-Δ effects) moments is treated as being equal to the product of the primary moment and a magnification factor (δ). The computed moments about each of the two principal axes are magnified separately, and the cross section is then proportioned for an axial load P and the magnified biaxial moments. The influence of the orthogonal moments on the magnifications, or on the stiffnesses, is ignored. The method is therefore unconservative.

A very conservative alternative is available. The ACI and other codes permit use of an 'equivalent circular compression member' for design purposes. The equivalent circular cross section is to have '... a diameter equal to the least lateral dimension of the actual shape', with the required reinforcement based on that dimension. This makes strength design as an equivalent uniaxially bent member simple, though frequently overly conservative, since in the case of circular cross sections the orthogonal moments can be added vectorially to produce a single resultant moment. The simplification is not applicable to the determination of the stiffness of members. The actual cross section must be used for that purpose.

The 1983 ACI Building Code provisions for slenderness evaluation of uniaxially bent reinforced concrete compression members recommend two alternate methods of stability analysis: the 'moment magnification method' and the 'second-order analysis'. Both methods will be described below, since the Code does specify their use in biaxial bending. The provisions of other international codes differ slightly from those adopted by ACI. However, the ACI procedures can be easily modified so that the requirements of the other codes are satisfied.

4.3.2.1 *Moment Magnification Method*

This method is applicable to columns with slenderness ratios (kl_u/r) less than 100. The slender column is designed as a short column subjected to the factored load P_u, from conventional frame analysis, and a simultaneous magnified factored moment M_c in each direction, defined by

$$M_c = \delta_b M_{2b} + \delta_s M_{2s} \qquad (4.6)$$

where δ_b is the magnification factor for the larger end moment caused by gravity loads M_{2b}, and δ_s is the magnification factor for the larger end moment due to lateral loads M_{2s}. The moments and their magnifiers are to be calculated separately in each direction.

The separate magnifier approach for the braced and unbraced portions of the frame action is based on the work of Ford *et al.* (1981). The magnification factors δ_b and δ_s are dependent on the cross section properties, the slenderness of the member, the restraints or applied moments at the ends, and the relative stiffness of the frame. Figures 4.4 and 4.5 outline the design procedure for evaluating the magnification factors and the magnified factored moments for biaxially bent slender columns in braced and unbraced frames. According to the ACI Commentary: 'A compression member may be assumed braced if located in a story in which the bracing elements (shearwalls, shear trusses, or other type of lateral bracing) have a total stiffness, resisting lateral

REINFORCED CONCRETE COLUMNS IN BIAXIAL BENDING

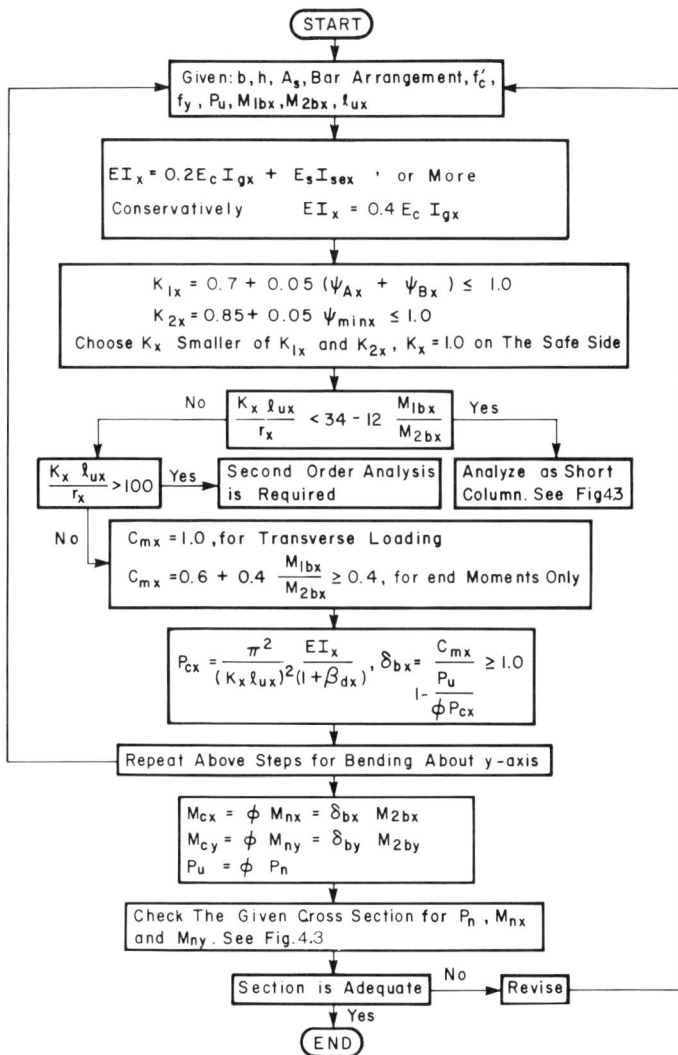

FIG. 4.4. Flowchart for design of biaxially bent slender columns in braced frames.

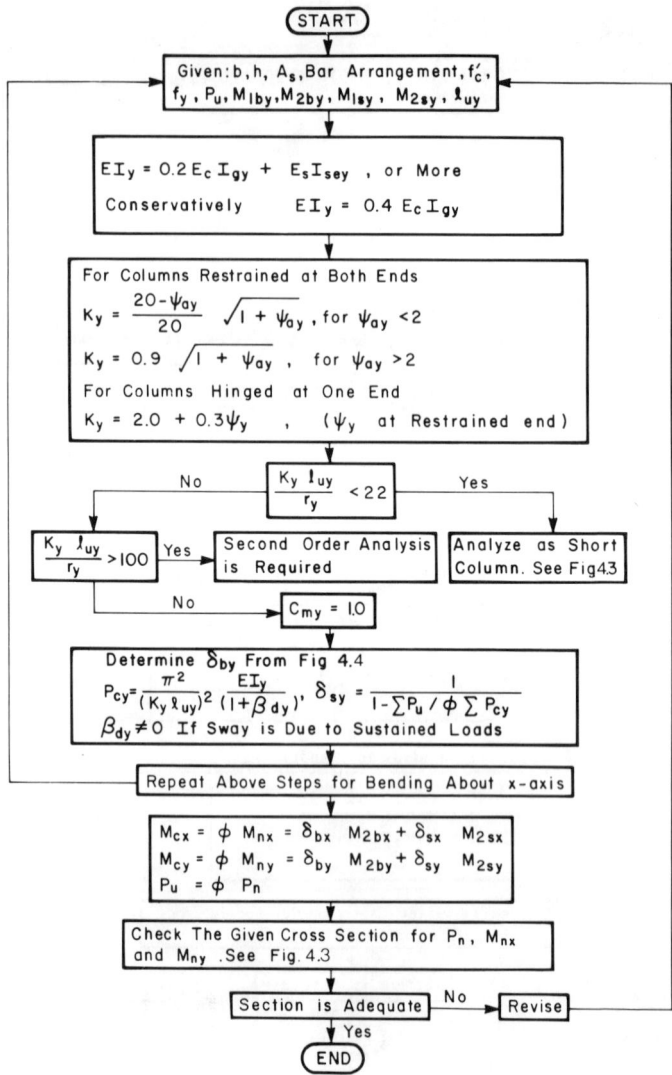

FIG. 4.5. Flowchart for design of biaxially bent slender columns in unbraced frames.

REINFORCED CONCRETE COLUMNS IN BIAXIAL BENDING   127

movement of the story, at least six times the sum of the stiffnesses of all the columns within the story'.

In lieu of a more accurate calculation, the ACI Code permits approximation of the column stiffness by the following expressions that include the effects of cracking and creep under long term loading, but not the effect of orthogonal moment:

$$EI = \frac{0 \cdot 2 E_c I_g + E_s I_{se}}{1 + \beta_d} \quad (4.7)$$

or use

$$EI = \frac{0 \cdot 4 E_c I_g}{1 + \beta_d} \quad (4.8)$$

Creep due to sustained loads is taken into account by introducing the factor $(1 + \beta_d)$ where β_d is the ratio of dead load moment to total load moment. Bazant and Tsubaki (1980) recommended that the factor $(1 + \beta_d)$ be replaced by $(1 + 1 \cdot 3\beta_d^2)$. Breen et al. (1972) recommended that the beam stiffness be based on the transformed cracked section, without specifying at what point along the beam the cross section is to be analyzed. ACI 318 is no more helpful, but it would seem reasonable to use the section adjacent to the column. The Commentary to ACI 318-83 suggests the use of $0 \cdot 5 EI_g$ for the beam and EI_g for the column when column kl_u/r is less than 60. I_g for the beam is not further defined in this section of the Commentary but, since the tests on which these recommendations were based were conducted with rectangular specimens, the flanges should presumably be disregarded for T shaped beams. See also Section 4.2.3 above, including the suggested reduction in column EI for frame analysis in the presence of orthogonal moments.

The ACI requires that the analysis of frame stability include the effects that accidental eccentricities or misalignment of the frame might produce. For design, the ACI requires that all columns, in braced and unbraced frames, be designed for at least a minimum eccentricity of $15 + 0 \cdot 03h$ mm ($0 \cdot 6 + 0 \cdot 03h$ in), about each axis. That means that either M_{2b} or M_{2s} must be greater than or equal to a moment calculated from this minimum eccentricity.

4.3.2.2 Second-order Analysis

When this more accurate method is used, the ACI Code does not limit design to members with kl_u/r less than 100. The second-order analysis must take into account the effect of deflections on moments and forces.

In general, economies in material quantities can be achieved by the use of the second-order analysis in sway or lightly braced frames since the moments are better approximations to the real moments than those of the moment magnification method (ACI 318-83 Commentary). A second-order analysis requires the use of sophisticated computer programs based on elastic frame analysis. Although the section stiffness of the concrete members is variable along the member and decreases as the moments are increased, the ACI 318-83 Commentary recommends the following stiffness approximations with uniaxial bending: (Note: use beam $I_g = \dfrac{b_w h^3}{12}$ throughout.)

$$\text{Column } EI = E_c I_g \left(0.2 + 1.2\rho g \frac{E_s}{E_c}\right) \quad (4.9)$$

$$\text{Beam} \quad EI = 0.5 E_c I_g \quad (4.10)$$

MacGregor and Hage (1977) recommended that the following be used:

$$\text{Column } EI = 0.8 E_c I_g \quad (4.11)$$

$$\text{Beam} \quad EI = 0.4 E_c I_g \quad (4.12)$$

Furlong (1981) presented a rational method of analysis for multistory concrete plane frames in which the following EI expressions were suggested:

Column at lowest level $\quad EI = 0.05 E_c b h^3 \quad (4.13)$

Column at other levels $\quad EI = 0.025 E_c b h^3 \quad (4.14)$

Beam $\quad EI = \dfrac{E_c b_w h^3}{12} \leqslant 0.5\, (EI_g \text{ of T section}) \quad (4.15)$

To account for accidental eccentricities or misalignment of the frame, Furlong recommended that the frame be analyzed as if it were 0.002 radian out of plumb. He did not discuss three dimensional frame action or biaxial bending in columns, but his recommended beam stiffnesses are larger and his recommended column stiffnesses are smaller than those of others. In *relative* terms they would therefore lead to moment distributions more like those influenced by orthogonal moments in the columns.

4.4 CONCLUSIONS AND FUTURE RESEARCH

When columns are subjected to biaxial bending, it is important to recognize that not only strength but also stiffness will be reduced in each direction due to orthogonal moments. The reduced stiffnesses will be obtained reasonably accurately by the method of Section 4.3.1. If insufficient time and computational resources can be made available to permit the more accurate analysis, the frame analysis can be carried out with the approximate stiffnesses suggested in Section 4.3.2. However, since those approximations were devised for uniaxially bent members, it would seem necessary to reduce the approximate column stiffnesses in proportion to the reduction in bending strength caused by the orthogonal moments.

The strength of biaxially bent columns can be obtained accurately by combining the methods of Sections 4.2.1 and 4.3.1. Figures 4.3, 4.4 and 4.5 provide flowcharts for obtaining the strength approximately. Repeated cycles of design and analysis will usually be required to obtain satisfactory and economical columns in a structure.

From the foregoing it may be concluded that the design of reinforced concrete columns subjected to simultaneous axial load and biaxial bending is a burdensome task. Fortunately, the economics of concrete construction usually dictate the column size more rigidly than strength requirements do. Thus, in order to achieve maximum repetition of formwork, many columns in a structure are made the same size. The need to reduce labor costs and to prevent congestion at column–beam intersections leads to simple and standard layouts of reinforcement, with much use of large bars and/or bundled reinforcement. The number of different cross sections encountered in a particular structural frame may therefore be relatively small.

Considerable savings in design effort can be achieved by choosing column dimensions which will be sufficiently large that, under the provisions of the governing code, slenderness need not be considered in the design. With some care in detailing, this can provide the additional benefit of preventing hinges from forming in the columns at collapse loads. Frames in which hinges form only in the beams will inherently be much more ductile than ones in which column hinges form. Use of larger columns will normally also mean use of fewer columns, with attendant savings in forming and increased flexibility afforded the owner and the architect in interior space layout.

Researchers could best help the designer at this time by conducting long term tests of short columns of realistic size to validate some of the theoretical calculations that have been carried out on the effects of creep, shrinkage and orthogonal moments on the strength and stiffness of a column. For this purpose representative service axial load and service biaxial moments should be applied, and held in place for perhaps a year, after which load combinations close to but not quite equal to the anticipated collapse loads should be applied and varied to provide stiffness information. After that, collapse loads can be used to determine remaining strength.

After experience has been gained with the short columns, the tests should be repeated on slender columns. Theoretical correlation must be to numerical models of the type described in Section 4.3.1, and given the variability of concrete, a number of different mixes and strengths will have to be looked at. Once validated, the numerical models can be used to provide reasonably simple approximations to the stiffness under any conditions.

It should be noted that similar long term stiffness tests are also needed for continuous beams, and especially ones cast integrally with floor slabs. Reports of stiffness tests of members kept under service load for long periods of time and then subjected to catastrophic overloading are conspicuous by their absence from the literature. And yet that condition is the one for which all 'ultimate' load design is carried out.

REFERENCES

AAS-JAKOBSEN, A. (1964) Biaxial eccentricities in ultimate load design. *ACI Journal,* Proc. **61**, March, 293-315.

ACI COMMITTEE 318 (1983) *Building Code Requirements for Reinforced Concrete,* American Concrete Institute, Detroit.

ACI COMMITTEE 318 (1983) *Commentary to Building Code Requirements for Reinforced Concrete,* American Concrete Institute, Detroit.

AL-NOURY, S. I. and CHEN, C. W. (1982) Finite segment method for biaxially loaded RC columns. *Journal of the Structural Division, ASCE,* **108**, April, 780-99.

AU, T. (1958) Ultimate strength design of rectangular concrete members subject to unsymmetrical bending. *ACI Journal,* Proc. **54**, February, 657-74.

BAZANT, Z. P. and TSUBAKI, T. (1980) Nonlinear creep buckling of reinforced concrete columns. *Journal of the Structural Division, ASCE,* **106**, November, 2235-57.

BREEN, J. E., MACGREGOR, J. G. and PFRANG, E. O. (1972) Determination of effective length factors for slender concrete columns. *ACI Journal,* Proc. **69**, November, 669-72.

BRESLER, B. (1960) Design criteria for reinforced columns under axial load and

biaxial bending. *ACI Journal,* Proc. **57**, November, 481–90. Discussion **57**, June 1961, 1621–38.
CHEN, W. F. and SHORAKA, M. T. (1975) Tangent stiffness method for biaxial bending of reinforced concrete columns. *IABSE,* **35-I**, 23–44.
CZERNIAK, E. (1962) Analytical approach to biaxial eccentricity. *Journal of the Structural Division, ASCE,* **88**, August, 105–58.
FORD, J. S., CHANG, D. C. and BREEN, J. E. (1981) Design indications from tests of unbraced multipanel concrete frames. *Concrete International, ACI,* **3**, March, 37–47.
FURLONG, R. W. (1961) Ultimate strength of square columns under biaxially eccentric loads. *ACI Journal,* Proc. **57**, March, 1129–40.
FURLONG, R. W. (1979) Concrete columns under biaxially eccentric thrust. *ACI Journal,* Proc. **76**, October, 1093–1118. Discussion **77**, August 1980, 295–6.
FURLONG, R. W. (1981) Rational analysis of multistory concrete structures. *Concrete International, ACI,* **3**, June, 29–35.
GESUND, H. (1967) Stress and moment distributions in three dimensional frames composed of non-prismatic members made of non-linear material. *Space Structures,* Blackwell Scientific Publications, Oxford, pp. 145–53.
GESUND, H. and VANDEVELDE, C. E. (1973) Stiffness of reinforced concrete columns in biaxial bending. *Planning and Design of Tall Buildings, Proceedings of 1972 ASCE–IABSE International Conference,* Vol. 3, pp. 455–9.
GOUWENS, A. J. (1975) Biaxial bending simplified. *Reinforced Concrete Columns,* SP-50, American Concrete Institute, Detroit, pp. 233–61.
MACGREGOR, J. G. and HAGE, S. E. (1977) Stability analysis and design of concrete frames. *Journal of the Structural Division, ASCE,* **103**, October, 1953–70.
MARIN, J. (1979) Design aids for L-shaped reinforced concrete columns. *ACI Journal,* Proc. **76**, November, 1197–216.
MEEK, J. L. (1963) Ultimate strength of columns with biaxially eccentric loads. *ACI Journal,* Proc. **60**, August, 1053–64.
PANNELL, F. N. (1960) The design of biaxially loaded columns by ultimate load methods. *Magazine of Concrete Research, London,* **35**, July, 99–108.
PANNELL, F. N. (1963) Failure surfaces for members in compression and biaxial bending. *ACI Journal,* Proc. **60**, January, 129–40.
PARME, A. L., NIEVES, J. M. and GOUWENS, A. J. (1966) Capacity of reinforced rectangular columns subject to biaxial bending. *ACI Journal,* Proc. **63**, September, 911–23.
PATEL, R. D. and GESUND, H. (1965) The ultimate strength of reinforced concrete columns under biaxially eccentric loading. *Indian Concrete Journal,* **39**, 250–7.
PINTO DE MAGALHAES, M. (1979) Biaxially loaded concrete sections. *Journal of Structural Engineering, ASCE,* **105**, December, 2639–56.
RAMAMURTHY, L. N. (1966) Investigation of the ultimate strength of square and rectangular columns under biaxially eccentric loads. *Symposium on Reinforced Concrete Columns,* SP-13, American Concrete Institute, Detroit, pp. 263–98.
RAMAMURTHY, L. N. and HAFEEZ KHAN, M. N. (1983) L-shaped column design for biaxial eccentricity. *Journal of Structural Engineering, ASCE,* **109**, August, 1903–17.

SHAH, M. J. and GESUND, H. (1972) The analysis of nonlinear three-dimensional frames. *Computers and Structures,* **2**, 943–54.

SHANMUGHASUNDARAM, N. (1977) Biaxial interaction exponent for concrete columns. *Journal of the Structural Division, ASCE,* **103**, December, 2295–306.

WANG, C. K. and SALMON, C. G. (1985) *Reinforced Concrete Design,* Fourth Edition, Harper and Row Publishers, New York.

WEBER, D. C. (1966) Ultimate strength design charts for columns with biaxial bending. *ACI Journal,* Proc. **63**, November, 1205–320.

Chapter 5

DESIGN OF CONCRETE STRUCTURES FOR TORSION

R. NARAYANAN

Department of Civil and Structural Engineering, University College, Cardiff, UK

SUMMARY

This chapter begins with a historical overview of various theories in torsion and then reviews the theoretical investigations reported in published literature on torsion in concrete structures. Methods of design for reinforced and prestressed concrete structures are outlined and the effect of interaction of torsion with coincident bending and shear is discussed. The chapter concludes with a summary of recent developments and the basis of design recommendations made in the CEB-FIP Model Code and in published literature.

NOTATION

A_s	Area of cross section of one leg of stirrup
A_1	Cross sectional area of all longitudinal bars
c	Coefficient relating the cube strength and tensile strength (eqn 5.6)
c'	Distance from centre of section to the centroid of longitudinal steel
f'_c	Cylinder strength of concrete (lb per sq.in)
f_{cu}	Cube strength of concrete
f_{ly}	Yield strength of longitudinal steel
f_r	Modulus of rupture of concrete
f_{sy}	Yield strength of steel used in stirrups
f_t	Tensile strength of concrete
G	Modulus of rigidity of concrete

h	Lever arm for bending
k_1, k_2	Coefficients accounting for the contributions to torsional strength due to concrete and steel (eqn 5.14)
M	Applied bending moment
M_u	Ultimate bending moment
M_{uc}	Ultimate bending capacity under combined loading
m	$(\rho_l f_{ly})/(\rho_s f_{sy})$
s_1	Spacing of stirrups
T	Applied torque
T_e^h	Elastic torsional strength of hollow section
T_p	Torsional capacity of plain concrete section
T_R	Torsion contributed by reinforcement
T_u	Ultimate torsional capacity
T_{uc}	Ultimate torsional capacity under combined loading
V	Applied shear force
V_u	Ultimate shear strength based on diagonal cracking in bending and shear
v_s	Diagonal tensile stress due to transverse shear
v_{sc}	Flexural shear stress carried by concrete when transverse shear and torsion act together
v_t	Shear stress due to torsion
v_{tc}	Torsional shear stress carried by concrete when transverse shear and torsion act together
x, y	Smaller and larger sides of the cross section
x_1, y_1	Smaller and larger centre-to-centre dimensions of rectangular stirrups
x', y'	Smaller and larger centre-to-centre dimensions of longitudinal steel
α, β	Coefficients in elastic torsional theory
σ_e	Magnitude of prestress after losses
ϕ	Angle of twist per unit length (mm^{-1})
ψ	Coefficient in Anderson's and Cowan's equations (eqn 5.11)

5.1 INTRODUCTION

Torsional moments are unavoidable in any practical beam, as it is impossible to ensure that the lines of action of all transverse forces always pass through the shear centre of the cross section. In the past,

DESIGN OF CONCRETE STRUCTURES FOR TORSION   135

torsion in concrete beams used to be regarded as a secondary effect; design calculations did not make explicit provisions for resisting torsion, relying instead on the reserve of strength implicit in the sturdy concrete sections provided. Present day designs using ultimate limit state methods and reduced load factors do not justify this procedure. Torsional moments induce shear stresses, which cause principal tensile stresses at 45° to the longitudinal axis of the beam; when these exceed the tensile strength of concrete, diagonal cracks are formed, which trigger collapse. Due to the brittle nature of concrete, sudden failure would occur immediately after the formation of the first crack, if no reinforcement were provided to resist torsion.

Few concrete members are subjected only to torsion; torsional moments are invariably encountered in conjunction with bending moments and shear forces; however, a knowledge of the behaviour of concrete structural components in pure torsion is basic to the analysis of such members subjected to the combined effect of bending, shear and torsion.

Before cracking occurs, the response of a beam to the application of a torsional moment is largely elastic and remains approximately the same whether the beam is made of plain, reinforced or prestressed concrete. Although the precracking stiffness is altered little, the ultimate strength and the post-cracking stiffness will depend significantly on the amount and disposition of reinforcement.

5.2 TORSION OF PLAIN CONCRETE MEMBERS

The torsional response of a plain concrete beam at early stages of loading can be closely predicted by employing the elastic theory due to St. Venant (1856), which accounts for the warping of non-circular cross sections subjected to torsion. For a rectangular section, the maximum torsional shear stress v_t and the angle of twist per unit length ϕ are given by the following equations:

$$v_t = \frac{T}{\alpha x^2 y} \quad (5.1)$$

$$\phi = \frac{T}{\beta x^3 y G} \quad (5.2)$$

where T is the applied torque, x and y are the smaller and larger sides of the cross section, G is the modulus of rigidity, and α and β are non-dimensional coefficients which depend on the value of (y/x).

St. Venant's theory was derived for a homogeneous material exhibiting linearly elastic behaviour. The maximum shear stress would occur at the middle of the larger side, and can be resolved into a compression and tension of magnitude equal to v_t, on 45° planes. As the tensile strength of concrete f_t is roughly a tenth of its compressive strength, it follows that the collapse due to torsional loading would be initiated when the diagonal tensile stress (due to torsion) equals the tensile strength of concrete f_t. Thus the maximum torque that can be sustained by a section ($T_{elastic}$) is obtained by replacing f_t for v_t in eqn (5.1) — see Bach and Graf, 1912 and Cowan, 1951.

$$T_{elastic} = \alpha x^2 y f_t \tag{5.3}$$

While the elastic theory described above is generally adequate to predict the initial loading response, tests have shown that the ultimate capacity of beams is consistently underpredicted by it. (Earlier investigators thought that the extra strength might be due to the plastic property developed by concrete near the ultimate loading stages — see Nylander, 1945.) If failure is assumed to occur when the maximum principal stress reaches the tensile strength f_t of concrete, the plastic failure torque T_{pl} can be estimated from the following equation (Nadai, 1950):

$$T_{pl} = \left(0.5 - \frac{x}{6y}\right) x^2 y f_t \tag{5.4}$$

Since concrete is a brittle material and does not exhibit any plastic characteristics, it is not surprising that the ultimate torques predicted by eqn (5.4) are unconservative (Zia, 1968).

In view of the uncertain nature of predictions obtained from the elastic and plastic theories, Hsu (1968a) carried out a large number of torsion tests to study the patterns of failure at the instant of collapse (using a high speed movie camera). He discovered that the failure process revealed a bending-type failure along a skew plane. This led to the following equation for the assessment of ultimate torque (T_{skew}) from the skew bending mechanism of failure:

$$T_{skew} = \frac{1}{3} x^2 y (0.85 f_r) \tag{5.5}$$

where f_r is the modulus of rupture of concrete.

Equation (5.5) is based on a failure criterion which was somewhat novel when first proposed: collapse is assumed to occur when the tensile stresses induced by the bending component of torque on the wider face reaches a suitably reduced value of modulus of rupture of concrete.

Hsu (1968a) has pointed out the similarity between eqns (5.3), (5.4) and (5.5), all having the same parameter x^2y; the only difference is in the non-dimensional coefficients and in the material constants. In the elastic and plastic theories, the material constant used is the direct tensile strength f_t, whereas in Hsu's theory a reduced modulus of rupture ($0.85 f_r$) is used. (To provide for the size effects observed in torsion tests, Hsu has proposed empirical equations for evaluating f_r.)

In most cases the direct tensile strength of concrete f_t is about $0.8 f_r$; the tensile strength (f_t, in N/mm^2) has also been related to the cube strength of concrete (f_{cu}, in N/mm^2) by an equation of the following type by many investigators (see Marshall, 1974):

$$f_t = c \sqrt{f_{cu}} \qquad (5.6)$$

where c is a coefficient having a dimension of $N^{0.5}$ mm^{-1}.

The value of c ranges from 0·37 to 0·45 depending on the material properties and the strength of the concrete mix.

Using the above, Narayanan and Kareem-Palanjian (1983) have shown that a lower bound estimate of torsional capacity of a plain concrete beam T_p can be obtained (in SI units) as

$$T_p = 0.13 x^2 y \sqrt{f_{cu}} \qquad (5.7)$$

The above equation has been derived by assuming that the cylinder strength of concrete f'_c can be taken as approximately $0.8 f_{cu}$. Equation (5.7) may be compared with the design equation proposed in ACI Specifications in FPS units:

$$T_p = \frac{1}{3} x^2 y (2.4 \sqrt{f'_c}) \qquad (5.8)$$

where f'_c is the cylinder strength in FPS units. After converting to SI units, the above equation can be rewritten in terms of f_{cu} (in SI units) as follows:

$$T_p = \frac{1}{3} x^2 y (0.18 \sqrt{f_{cu}}) \qquad (5.9)$$

A comparison of eqns (5.7) and (5.9) shows that eqn (5.9) is very conservative.

Marshall (1974) has suggested that a reliable estimate of the ultimate torsional strength of a plain concrete section could be obtained by substituting the split cylinder strength f_{sp} for f_t in eqn (5.4).

5.3 REINFORCED CONCRETE IN TORSION

As stated previously, the torque causing the first crack in a conventionally reinforced beam is very nearly the same as the ultimate torque of the corresponding beam made of plain concrete. Hence the cracking torque can be calculated by the methods discussed in the previous sections.

Early studies on plain concrete members subjected to torsion led to the provision of a 45° spiral reinforcement normal to the crack, which, in practice, proved to be expensive. The same objective can be achieved by a combination of longitudinal and transverse reinforcement, to withstand the horizontal and vertical components of the diagonal tensile forces due to torsion; hence, it is obvious that the provision of either longitudinal or transverse steel by itself would not significantly improve the torsional strength of concrete sections. The provision of web reinforcement together with longitudinal reinforcement introduces significant changes in the post-cracking behaviour of reinforced concrete sections subjected to torsion. The presence of the web reinforcement allows a redistribution of internal stresses after the formation of the first diagonal crack, resulting in an increased torsional capacity; collapse occurs at a significantly enhanced loading, after extensive additional cracking.

The first rational theory for reinforced concrete in torsion was developed by Rausch (1929), who used a space truss model in which the concrete is represented by 45° diagonal concrete struts resisting compression, the longitudinal and lateral bars carrying only tension. Only a hollow section of concrete is considered, with the core making no contribution to the ultimate torque. The resulting ultimate torsional strength T_u was derived as

$$T_u = 2\frac{x_1 y_1 A_s f_{sy}}{s_1} \qquad (5.10)$$

where A_s is the area of cross section of one leg of a stirrup, f_{sy} is the yield strength of steel used as stirrups, s_1 is the spacing of stirrups, and x_1 and y_1 are the smaller and larger centre-to-centre dimensions of the rectangular stirrup.

DESIGN OF CONCRETE STRUCTURES FOR TORSION

Anderson (1935) pointed out that the truss analogy, postulated above, assumes a uniform stress all along the reinforcement in a member subjected to torsion. In a rectangular cross section, the maximum stress occurs in the middle of the wider face and decreases to zero at the corners. To allow for this, he modified Rausch's equation by assuming that the contribution of concrete is equal to the torque capacity of the corresponding plain concrete based on elastic theory; he also introduced an efficiency factor (less than unity) to the contribution of ultimate torque by steel to the ultimate capacity; this was expressed as a function of the y/x ratio and the distribution of the reinforcement.

Cowan (1953) used a strain energy method, and modified Rausch's equation by an efficiency factor by equating the external work done by the applied torque to the strain energy stored in steel and concrete. Although his theory was based on a working stress method, he suggested that it could be extended to the evaluation of the ultimate torque of the reinforced concrete members:

$$T_u = \alpha x^2 y f_t + \frac{x_1 y_1 A_s f_{sy} \psi}{s_1} \qquad (5.11)$$

where ψ is a constant having a value of 1·33 and 1·6 according to Anderson and Cowan respectively.

More recently, a number of analytical approaches have been proposed to predict the ultimate strength of reinforced concrete members under torsion. These methods are primarily based on either the ultimate equilibrium method or space truss analogy.

The equilibrium approach was first presented by Lessig (1959) and later modified by Yudin (1962). This analysis was based on the equilibrium of forces across an assumed failure surface (Fig. 5.1). The theory was developed for beams subjected to combined bending and

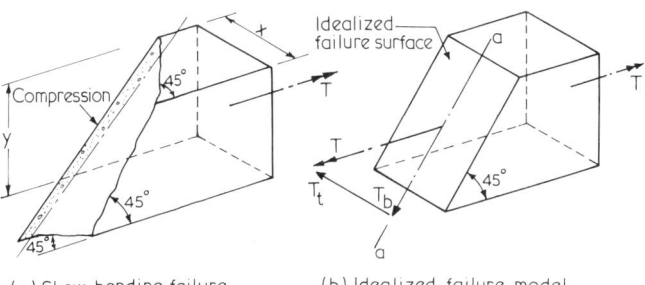

(a) Skew bending failure   (b) Idealized failure model

FIG. 5.1. Torsional cracking of plain concrete beam.

torsion, with pure torsion treated as a special case. The failure of the beam subjected to torsion or combined bending and torsion was taken as occurring on a warped surface bounded by tensile cracks on three faces and a compression zone on the fourth face, as shown in Fig. 5.1(a). The compression zone could be either on the top surface of the beam or on one of the vertical sides. As the ultimate torque was reached, it was assumed that the two segments of the beam rotated about a hinge located along a neutral axis. The magnitude of the limiting external forces causing the formation of a hinge along the neutral axis could be determined from components of the internal forces in the concrete and reinforcement which opposed the rotation about the hinge. The ultimate torsional strength was derived on the basis of two equilibrium conditions, namely

(a) the total sum of the moments of external and internal forces acting in the plane normal to the neutral axis, about the centre of gravity of the compression zone, is equal to zero;
(b) the components of all internal and external forces on the axis perpendicular to the plane of the compression zone are equal to zero.

The actual failure surface was determined by minimizing the resistance of the section. It was found that the theoretical minimum torsional resistance occurred when the neutral axis was parallel to either a shorter face or a longer one. This resulted in two modes of failure, namely Mode I and Mode II failure respectively (Fig. 5.2);

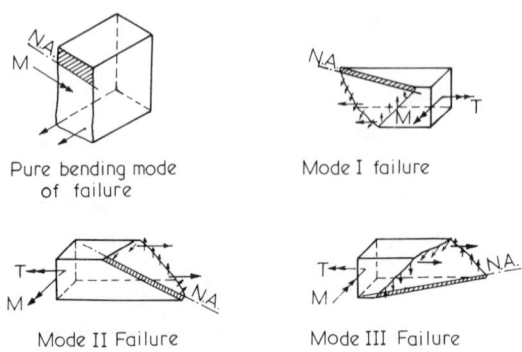

Different Modes of Failure

FIG. 5.2. Different failure surfaces and modes of failure.

DESIGN OF CONCRETE STRUCTURES FOR TORSION 141

Mode I failure has the compression zone near the top face whereas Mode II failure is characterized by compression near a side face. Assuming that all steel crossing the failure surface of a symmetrical section yielded, the ultimate torque was obtained from

$$T_u = 2x_1y_1 \sqrt{\frac{A_s f_{sy}}{s_1} \frac{A_1 f_{ly}}{2(x_1 + y_1)}} \qquad (5.12)$$

where A_1 is the cross sectional area of all longitudinal bars and f_{ly} is the yield strength of longitudinal steel.

Yudin (1962) pointed out that Lessig employed only two equilibrium conditions, viz. a force equilibrium and a moment equilibrium, which was insufficient to determine the volume ratio of the longitudinal steel and stirrups. To overcome this, he suggested using a third (moment) equilibrium equation so that the amount of longitudinal and transverse steel could be estimated. Assuming a failure surface cracked at 45° on each surface, and neglecting the resistance of the compression zone and the compression steel, Yudin derived an equation identical to Rausch's equation as shown in Table 5.1.

Hsu (1968b) reported that the Lessig-Yudin theory over-estimated the torsional resistance and did not account for the dowel action of longitudinal corner bars. He observed that the diagonal cracks on the shorter face formed at an angle much less than 45°. Moreover the shorter legs of stirrups usually had only a small tensile stress at the ultimate stage. To allow for this, he assumed an idealized failure surface, similar to plain concrete section, and developed an equation for the ultimate torsional strength of reinforced concrete beams based on a semi-empirical approach. The idealized skew surface was proposed on a plane perpendicular to the wider face and inclined at 45° to the axis of the beam; the proposed failure surface did not intersect the shorter legs of the stirrups; the method took into account the shear resistance of the concrete compression zone and the dowel forces of the longitudinal steel. Hsu's expression for the ultimate torque of a reinforced concrete rectangular member with equal longitudinal steel at the top and bottom faces is given in Table 5.1. The ACI Code (1971) generally follows Hsu's approach.

Another rational analysis for ultimate torsional strength of reinforced concrete members is based on space truss analogy, referred to already. The elementary theory by Rausch (1929) required equal volumes of longitudinal and transverse steel. This 45° truss model was later generalized for members in torsion and bending (with pure torsion

TABLE 5.1
ULTIMATE TORQUE EXPRESSIONS DERIVED BY PREVIOUS INVESTIGATORS

Investigator	Basis	Expression for ultimate torque (in SI units)		Limitations	Inapplicable for
		Contribution from concrete	Contribution from reinforcement		
Rausch (1929) (also Yudin, 1962)	Truss analogy (equilibrium)	—	$\dfrac{2x_1 y_1 A_s f_{sy}}{s_1}$	—	—
Anderson (1937)	Semi-empirical	$ax^2 y f_t$	$\dfrac{1 \cdot 33 x_1 y_1 A_s f_{sy}}{s_1}$	—	—
Cowan (1953)	Strain energy	$ax^2 y f_t$	$\dfrac{1 \cdot 6 x_1 y_1 A_s f_{sy}}{s_1}$	—	—
Lessig (1959)	Equilibrium	—	$2x_1 y_1 \sqrt{\dfrac{A_s f_{sy}}{s_1} \dfrac{A_l f_{ly}}{2(x_1 + y_1)}}$	—	—
Hsu (1968b)	Skew bending (equilibrium)	$\dfrac{0 \cdot 9}{\sqrt{x}} x^2 y \sqrt{f_{cu}} +$	$\left(0 \cdot 66 \dfrac{\rho_l}{\rho_s} + 0 \cdot 33 \dfrac{y_1}{x_1}\right) \dfrac{x_1 y_1 A_s f_{sy}}{s_1}$	$1 \cdot 5 < \dfrac{\rho_l}{\rho_s} < 0 \cdot 7$ $\dfrac{y_1}{x_1} \leqslant 2 \cdot 6$	$s_1 > \dfrac{y_1}{2}$ Unsymmetrical longitudinal reinforcement

DESIGN OF CONCRETE STRUCTURES FOR TORSION

Author	Method	Formula		
Lampert and Thurlimann (1968)	(Space truss analogy)	—	—	$\sqrt{2} < \dfrac{\rho_l f_{ly}}{\rho_s f_{sy}} < \dfrac{1}{\sqrt{2}}$
Kuyt (1971)	Space truss analogy	$2x_1 y_1 \sqrt{\dfrac{A_s f_{sy}}{s_1} \dfrac{A_l f_{ly}}{2(x_1+y_1)}}$	—	—
ACI Code (1971)	Skew bending	$0{\cdot}059\, x^2 y \sqrt{f_{cu}} + \left(0{\cdot}66 + \dfrac{0{\cdot}33\, y_1}{x_1}\right) \dfrac{x_1 y_1 A_s f_{sy}}{s_1}$	$\dfrac{y_1}{x_1} \leqslant 2{\cdot}6$	Expressions are the same as Lampert and Thurlimann given above $(x_1, y_1$ measured from centre to centre of stirrups) $s_1 < \dfrac{x_1 + y_1}{4}$ Unsymmetrical longitudinal reinforcement
Victor and Muthukrishnan (1973)	Semi-empirical	$(0{\cdot}13\, x^2 y \sqrt{f_{cu}}) + k_2 \dfrac{x_1 y_1 A_s f_{sy}}{s_1}$	—	—

treated as a special case) by Lampert and Thurlimann (1968); this was later modified by Kuyt (1971), based on the behaviour of solid and hollow sections. Test results by Hsu (1968b) showed that hollow sections had a smaller cracking torque than the equivalent solid sections of the same size and the same concrete quality but there was little difference in the ultimate torque. Hence, the truss theory postulated that the solid reinforced concrete section after cracking could be replaced by a thin walled hollow section in which the reinforcement cage became effective, and provided the total resisting torque. In each shear wall, the concrete between consecutive diagonal cracks was assumed to provide the compressive diagonal with the longitudinal bars acting as stringers and stirrups as posts (Fig. 5.3). By studying the equilibrium of this truss model, an expression for torsional capacity similar to eqn (5.12) was derived for symmetrically reinforced sections. Kuyt pointed out that eqn (5.12) would be valid only if the longitudinal steel was distributed evenly along the perimeter, and derived the expression for ultimate torque for the case of the four corner bars to be

$$T_u = 2x_1 y_1 \sqrt{\frac{x_1 + y_1}{2y_1} \frac{A_1 f_{ly}}{2(x_1 + y_1)} \frac{A_s f_{sy}}{s_1}} \qquad (5.13)$$

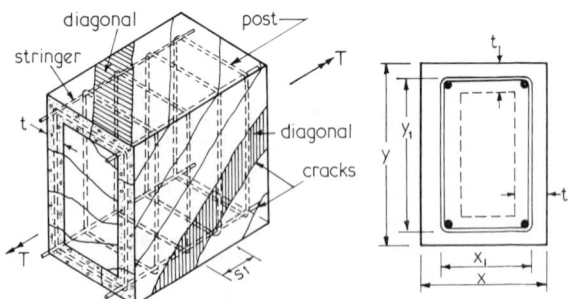

FIG. 5.3. The space truss model.

This idea was further developed by Mitchell and Collins (1974) who modified the truss model, which they called the 'diagonal compression field theory'. Their model is capable of satisfying both equilibrium and compatibility conditions.

Victor and Muthukrishnan (1973) showed that for a given volume fraction of stirrups, higher torsional strength was obtained for closer spacing of the stirrups assuming that adequate longitudinal steel had

DESIGN OF CONCRETE STRUCTURES FOR TORSION   145

also been provided. The torsional strength of a beam with a symmetrical arrangement of longitudinal steel was found to be greater than that obtained with the same amount of steel arranged unsymmetrically (Pandit, 1970). Moreover, if all the longitudinal steel was provided at the bottom, the torsional strength was found to be almost the same as that of the corresponding plain concrete section. These characteristics are not explained by any of the theoretical methods detailed above. It is nevertheless important to be able to predict the ultimate torque under pure torsion, even though such a loading condition is seldom met in practice.

To overcome the above shortcomings, semi-empirical approaches have been proposed by Rajagopalan *et al.* (1971), Pandit (1970), Victor and Muthukrishnan (1973) and Chakraborty (1979). These methods account for the effectiveness of the reinforcement as well as the contribution of the concrete to the ultimate torsional strength; a summary of these methods along with the more rational methods is given in Table 5.1.

More recently, Pandit and Venkappa (1980) suggested a common form for computing the ultimate torque, which encompassed the various theories and semi-empirical relations suggested by previous investigators.

$$T_u = k_1 T_p + k_2 T_R \qquad (5.14)$$

where k_1 and k_2 are constants which account for the contribution of concrete and reinforcement respectively to the ultimate torsional strength of reinforced concrete sections, and T_R is the contribution of torsional reinforcement given by

$$T_R = \frac{x_1 y_1 A_s f_{sy}}{s_1} \qquad (5.15)$$

They compared the ultimate torques predicted by the methods proposed by various investigators with the observed ultimate torques from 100 tests reported from nine published reports, and showed that the semi-empirical method proposed by Victor and Muthukrishnan (1973) gave the best correlation with test results. The predictions of the ultimate torques obtained were safe as long as the beams were provided with any reasonably practical spacing of transverse reinforcement in all the following cases:

(a) when the beams were under-reinforced or over-reinforced;

(b) when longitudinal reinforcement was placed symmetrically or unsymmetrically;
(c) when the cross section was rectangular, L-shaped or T-shaped.

The expression proposed by them for rectangular beams is given below:

$$T_u = T_p + k_2 T_R$$

$$T_u = 0 \cdot 13 x^2 y \sqrt{f_{cu}} + k_2 \frac{x_1 y_1 A_s f_{sy}}{s_1} \qquad (5.16)$$

where

$$k_2 = \left[1 - \frac{2c'}{y} \frac{f_{lyb}}{f_{lyt}}\right]\left[0 \cdot 2m + \sqrt{m}\left(0 \cdot 45 \frac{y_1}{x_1} - \frac{s_1}{x_1 + y_1}\right)\right] \qquad (5.17)$$

$$m = \frac{\rho_l f_{ly}}{\rho_s f_{sy}} \quad \text{and} \quad f_{ly} \leqslant f_{sy}$$

c' is the distance from the centre of the section to the centre of gravity of total longitudinal steel, f_{lyb} and f_{lyt} are yield stresses of the bottom and top longitudinal steel respectively, and ρ_l and ρ_s are the volume fractions of longitudinal steel and stirrups respectively.

5.4 PRESTRESSED CONCRETE BEAMS SUBJECTED TO TORSION

Prestressed concrete beams exhibit brittle characteristics and are similar to plain concrete beams unless additional torsional reinforcement is provided. Subjecting concrete members to axial prestress results in a substantial increase in the torsional capacities. However, the addition of prestress alone will not add any post-cracking ductility to the member. In prestressed concrete members subjected to torsion, there exists a state of biaxial stress; cracking occurs when the combined stress produced by the torsional shear and prestress exceeds the concrete tensile strength as defined by an appropriate failure criterion.

Tests on prestressed concrete members in pure torsion were first carried out by Nylander (1945), who reported on 15 uniformly prestressed rectangular beams. He observed that the prestressed members reached their ultimate torques immediately after the formation of the first crack and established that the torsional strength of the member was increased significantly by axial compression. He also found that the direction of the failure cracks depended on the level of prestressing. Later Cowan and Armstrong (1955) found that the greatest increase in torsional strength was obtained by uniform prestressing. Cowan (1953), taking account of biaxial stresses, suggested a dual criterion of failure for concrete, combining Rankine's maximum stress theory with Mohr's generalized shear theory, as shown in Fig. 5.4. His suggestion was based on the observation of two distinct modes of failure of concrete, viz. cleavage and shear fracture (crushing). He simplified Mohr's stress envelope by replacing it by two straight lines, i.e. a vertical line and a straight line inclined at an angle of 37° to the normal axis, which was assumed to be the angle of internal friction of concrete aggregate. According to Cowan, a crushing failure was generally caused by diagonal shear due to compression and was characterized by the formation of debris. A cleavage failure could be easily recognized by the clean appearance of the fracture. If the beam was suitably reinforced, failure in direct tension due to bending would be prevented, and a

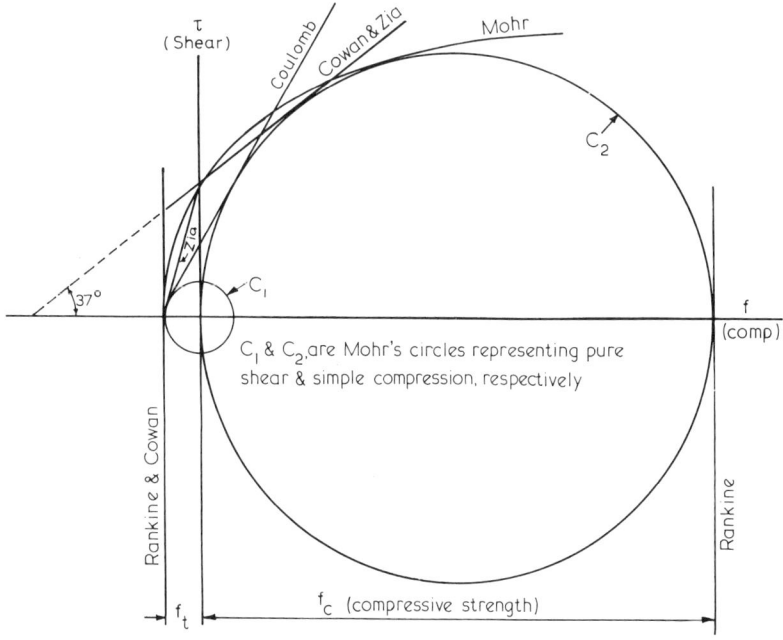

FIG. 5.4. Failure theories.

cleavage failure would normally be the result of diagonal tension due to shear. Assuming that the failure was determined by the criterion of a constant maximum tension, the maximum shear stress in a prestressed beam due to torsion f_{tu} was computed from

$$f_{tu} = \sqrt{f_t^2 + f_t \sigma_e} \tag{5.18}$$

where σ_e is the magnitude of prestress after all losses. Hence, the maximum twisting moment T_u of prestressed concrete beams is given by

$$T_u = \alpha x^2 y f_t \sqrt{1 + \frac{\sigma_e}{f_t}} \tag{5.19}$$

where f_t is the maximum permissible stress for the concrete in diagonal tension.

Based on his experimental studies, Humphreys (1957) suggested that the maximum torsional shear be computed from $(f_t + 0.3\sigma_e)$ and proposed the following equations for predicting the ultimate torque of bonded prestressed rectangular concrete sections using elastic theory.

(a) For centrally prestressed sections:

$$T_u = \alpha x^2 y (f_t + 0.3\sigma_e) \tag{5.20}$$

where the effective prestress $\sigma_e \leqslant 0.5 f_{cu}$.

(b) For eccentric prestressing the lower value obtained from the two equations given below:

$$\left. \begin{array}{l} T_u = \alpha k' x^2 y (f_t + 0.3\sigma_{e1}) \\ T_u = \alpha x^2 y (f_t + 0.3\sigma_{e2}) \end{array} \right\} \text{ whichever is lower} \tag{5.21}$$

where σ_{e1} and σ_{e2} are the minimum effective compressive stresses present on the shorter and longer sides of the specimens respectively. k' is a constant depending on the y/x ratio, which accounts for the difference in the prestress at the centre of longer and shorter faces respectively.

Zia (1961) tested 68 prestressed and plain concrete members consisting of rectangular, T- and I-sections under pure torsion. Some of the specimens also contained web reinforcement in the form of ties. He examined various theories of failure for concrete under combined stresses and concluded that Rankine's maximum stress theory over-estimated the torsional strength of prestressed concrete and Coulomb's internal friction theory underestimated it. Cowan's theory overestimated the strength if the failure was predominantly of the cleavage type, but conservatively predicted the strength if the failure was primarily of the fracture type. Zia proposed a modification to Cowan's theory to account for the torsional failure of concrete from distinct cleavage to fracture, by drawing a line from the point of intersection of the Cowan envelope with the torsion shear (τ) axis and tangent to Circle C_1 (as shown in Fig. 5.4). His test results indicated that the ultimate torque of prestressed members of rectangular and T-sections without transverse reinforcement was the same as the cracking torque calculated according to St. Venant's torsion theory and the failure criterion proposed by him. However, the test results for I-sections exceeded the predicted strengths. Zia's test results showed that the direction of cracks depended on the magnitude and eccentricity of the prestress; the larger the prestress, the smaller the angle of inclination of the crack. For the eccentrically prestressed members, the cracks usually formed well above the centre of gravity of the prestressing steel and at an angle to the axis of the member; the cracks then propagated toward the prestressing steel with decreasing slope. Zia concluded that the use of web reinforcement had no noticeable effect on the elastic torsional behaviour of a prestressed member; however, the ductility of the member would be improved, and the ultimate strength would equal the sum of cracking torque of the member and torque resisted by the web steel.

Swamy (1962) used Cowan's approach to predict the torsional strength of hollow concrete sections, prestressed uniformly, as follows:

$$T_u = T_e^h \sqrt{1 + \frac{\sigma_e}{f_t}} \qquad (5.22)$$

where T_e^h is the elastic torsional strength of a plain concrete hollow section.

Hsu (1984) extended the skew bending theory of plain concrete to uniformly prestressed members by using the following modification:

$$T_u = T_p \sqrt{1 + \frac{\sigma_e}{f_t}} \qquad (5.23)$$

By using $f_t = f'_c/10$ this can be re-written as

$$T_u = T_p \sqrt{1 + \frac{10\sigma_e}{f'_c}} \qquad (5.24)$$

In the above, T_p was calculated using the following empirical relation (in imperial units):

$$T_p = 6(x^2 + 10)y\sqrt[3]{f'_c} \qquad (5.25)$$

Hsu (1968c) analyzed the published test results of many investigators on prestressed concrete beams and the corresponding plain concrete beams and showed that the multiplying factor $\sqrt{1 + 10\,\sigma_e/f'_c}$ applied at least up to an upper limit of $\sigma_e/f'_c = 0.45$. This was not a serious restriction as most practical concretes would satisfy this requirement.

Hsu and Kemp (1969) equated $0.85 f_r$ in eqn (5.5) to $6\sqrt{f'_c}$ (in imperial units) and proposed the following expression for prestressed beams without web reinforcement (in imperial units):

$$T_u = \frac{1}{3}x^2 y\,(6\sqrt{f'_c})\sqrt{1 + \frac{10\,\sigma_e}{f'_c}} \qquad (5.26)$$

In terms of the cube strength f_{cu} (in SI units) the above equation can be re-written as

$$T_u = \frac{1}{3}x^2 y\,(0.444\sqrt{f_{cu}})\sqrt{1 + 12.5\frac{\sigma_e}{f_{cu}}} \qquad (5.27)$$

Hsu and Kemp suggested that the ACI formula given in Table 5.1 be extended for prestressed concrete members containing web reinforcement as given below (in imperial units):

$$T_u = 0.4\left(\frac{1}{3}x^2 y\right)6\sqrt{f'_c}\left[2.5\sqrt{1 + \frac{10\,\sigma_e}{f'_c}} - 1.5\right]$$

$$+ \left(0.66 + 0.33\frac{y_1}{x_1}\right)\left(\frac{x_1 y_1 A_s f_{sy}}{s_1}\right) \qquad (5.28)$$

In terms of the cube strength f_{cu} (in SI units) the equation can be written as

$$T_u = 0.4\left[\frac{1}{3}x^2 y\right]\left[0.444\sqrt{f_{cu}}\right]\left[2.5\sqrt{1 + \frac{12.5\,\sigma_e}{f_{cu}}} - 1.5\right]$$

$$+ \left(0.66 + \frac{0.33\,y_1}{x_1}\right)\left(\frac{x_1 y_1 A_s f_{sy}}{s_1}\right) \qquad (5.29)$$

DESIGN OF CONCRETE STRUCTURES FOR TORSION   151

Bishara and Peir (1968) found that the application of an axial load or uniform prestress to a member subjected to pure torsion increased its torsional capacity up to a certain transformation point where the applied stress was about $0.65 f'_c$. Beyond this limit, the torsional capacity decreased with any further increase of axial load. They proposed the following semi-empirical equation to predict the ultimate torque capacity of axially loaded members under pure torsion on the basis of plastic theory, when σ_e is less than $0.65 f'_c$:

$$T_u = \left[\left(0.5 - \frac{x}{6y}\right) x^2 y \, v'_c + \Omega \frac{x_1 y_1 A_s f_{sy}}{s_1}\right] \sqrt{1 + \frac{12 \sigma_e}{f'_c}} \quad (5.30)$$

$$\text{where } \Omega = 0.66 m \frac{x' y'}{x_1 y_1} + 0.33 \frac{y_1}{x_1}, \quad v'_c = \frac{f'_c \cdot f_t}{(f'_c + f_t)} \quad (5.31)$$

and x' and y' are the smaller and larger centre-to-centre dimensions of longitudinal steel respectively.

Evans and Khalil (1970) tested uniformly and eccentrically prestressed concrete beams, with or without web reinforcement. They concluded that beams eccentrically prestressed with zero stress at the top and having no web reinforcement, when subjected to pure torsion, failed immediately after the formation of the first crack. The failure was not governed by the stress at the top face but by the stress at the middle of the longer faces. They also observed that in prestressed beams reinforced with stirrups, and subjected to pure torsion, cracking started at the top face of the beam where the prestress was zero. The inclination of the crack was 45° with the beam axis. As the torque increased, cracks travelled downwards on to the vertical sides of the beam. Failure was of the cleavage type and some crushing occurred at the face of the beam where the maximum prestress acted.

Mukherjee and Warwaruk (1971) concluded from their torsion tests that the provision of reinforcement in the form of rectangular stirrups and longitudinal bars at the corners prevented a brittle type failure in prestressed concrete beams; the rotation capacity was substantially increased with the accompaniment of some increase in ultimate strength in pure torsion. They also found that an increase in prestress caused a corresponding increase in the ultimate strength, but reduced the rotation capacity. Eccentricity of prestressing did not reduce the ultimate torque; in fact, a slight increase was observed by them.

Ganga Rao and Zia (1973) tested six rectangular prestressed beams under pure torsion. The beams were provided with longitudinal and transverse steel and were eccentrically or centrally prestressed. They observed that the beams exhibited considerable ductility at failure which improved when the stirrup spacing was reduced or when prestressing was eccentric; the use of longitudinal mild steel also improved ductility. The initial crack developed at the centre of the wide face for centrally prestressed beams, and at the top for eccentrically prestressed beams. The cracking moment capacity reduced slightly for eccentrically prestressed beams and the ultimate torque increased slightly as compared with the centrally prestressed specimens. The addition of longitudinal mild steel did not result in a significant increase in pure torsional strength.

Zia and McGee (1974) on the basis of extensive studies recommended that eqn (5.26) should be modified as follows (in imperial units):

$$T_u = \left[\frac{0.35}{(0.75 + y/x)}\right] x^2 y \, (6\sqrt{f'_c}) \sqrt{1 + \frac{10\sigma_e}{f'_c}} \qquad (5.32)$$

Many other experimental studies have been reported in published literature (e.g. Pandit and Mawal, 1973; Iyengar et al., 1974). From these, it may be concluded that the torsional strength of plain prestressed concrete sections (without web reinforcement) can be expressed by a relationship of the following type:

$$T_u = T_p \sqrt{1 + \frac{\sigma_e}{f_t}} \qquad (5.33)$$

where T_p is calculated by the methods discussed previously.

The torsional strength of prestressed concrete sections provided with web reinforcement can be expressed using a relation of the type

$$T_u = k_1 T_p \sqrt{1 + \frac{\sigma_e}{f_t}} + k_2 T_R \qquad (5.34)$$

where T_R, k_1 and k_2 are also calculated by the methods discussed previously.

5.5 BEAMS SUBJECTED TO COMBINED BENDING AND TORSION

In plain concrete sections, the first crack leads to instant failure both in bending and torsion. Based on the principal tensile stress criterion,

DESIGN OF CONCRETE STRUCTURES FOR TORSION   153

Cowan (1953) has derived an interaction curve for combined bending and torsion of plain concrete sections at ultimate stages:

$$\left[\frac{T}{T_p}\right]^2 + \left[\frac{M}{M_u}\right] = 1 \qquad (5.35)$$

where T and M are the applied torque and bending moment respectively, and T_p and M_u are the ultimate torque and moment capacities of the section respectively.

Plain concrete members are never used in practice to resist bending and torsional moments, and hence their behaviour is not of any great practical significance. The strength and behaviour of reinforced concrete sections under combined bending and torsion are greatly influenced by the shape of the cross section, the amount and distribution of the steel, and bending moment to torque ratio. For members with longitudinal reinforcement only, previous investigators have proposed many different types of interaction curves based on experimental results (Zia, 1970). These investigations reveal that, in general, the presence of a small amount of bending moment increases the torsional strength slightly, particularly for T-beams. On the other hand, for small amounts of torsion, there is almost no reduction in the bending moment capacity.

An interaction diagram based on a combined maximum stress theory and internal friction theory of failure for reinforced concrete beams was obtained by Cowan (1953). If torsion predominates, failure is governed by the maximum stress theory where the bending stress is determined on the basis of a transformed uncracked section. On the other hand, the internal friction theory is the controlling factor if bending predominates, in which case the bending stress is determined on the basis of a cracked section.

Tests by Cowan and Armstrong (1955) showed good correlation with the theory. Both the theoretical and experimental results showed that a small amount of bending will improve the torque capacity but a slight reduction in bending capacity results by including a small amount of torque. In other words, when torsional mode of failure occurs, the presence of bending slightly increases the torsional strength; this increase continues till the point where failure changes to the bending type. When the bending mode of failure predominates, the presence of torsion would decrease the ultimate bending moment.

A number of ultimate strength theories have been proposed for beams with web reinforcement under combined bending and torsion.

These are also based on the two approaches discussed previously, viz. (a) the skew bending or the ultimate equilibrium method and (b) the space truss method. The ultimate equilibrium method originally proposed by Lessig (1959) has been subsequently modified and extended by Goode and Helmy (1968), Collins *et al.* (1968) and McMullen and Warwaruk (1970). The major difference between the various approaches is the shape of failure surface, the direction of the axes about which the beam rotates, and the location of the compression zone. While the earlier investigators considered two modes of failure, with the compression zone assumed to form either on the top surface or on the vertical side, Goode and Helmy, Collins *et al.* and McMullen and Warwaruk have included a third mode of failure where the compression zone might form at the bottom of the beam (see Fig. 5.2).

To explain the failure surface of a reinforced concrete member in combined bending and torsion, Warner *et al.* (1977) considered the behaviour of a beam in pure bending first and then examined the effect of torsion superimposed on the bending. If the beam is under-reinforced and subjected to pure bending, the bottom steel yields before the concrete crushes. The failure surface is simply a normal cross section of the beam and does not intersect any stirrups. The effect of a small addition of twisting moment to the beam will skew the resultant moment vector at failure, thus causing a skewed failure surface. The compression zone changes direction from across the beam to an angle to the cross section (see Fig. 5.3). The skewed failure surface intersects some of the web reinforcement as well as the longitudinal steel. Such a failure is designated as Mode I failure, which is a modification of bending failure. In this mode, the beam carries a substantial bending in addition to some torsion with the compression zone at the top skewed to the beam axis. When the beam is subjected primarily to torsion with some bending, the behaviour depends on the dimensions of the cross section. If the section is rectangular ($x < y$) with equal top and bottom steel, a lateral bending type of failure may occur in which the compression zone will form on one of the sides of the beam. This type of failure is designated as Mode II. On the other hand, if the section is square and is symmetrically reinforced, a Mode I failure may still occur. For a beam reinforced with more steel at the bottom than at the top, and subjected to a high torque to moment ratio, the compression zone might form at the bottom of the beam. Such a failure is called Mode III failure; in this case also, when high bending moment is present, it might control the failure and force a Mode I failure. In pure torsion, the beam will fail about its weakest bending axis (laterally for rectangular beams and in Mode III if there is a deficiency of top steel). A small bending moment may not be sufficient to convert the type of failure away from the mode for pure torsion.

A rational theory based on the space truss analogy for assessing the strength of sections in combined bending and torsion was presented by Lampert and Collins (1972) following the application of the method to combined loading by Lampert and Thurlimann (1971). It has been observed that if torsion dominates in beams with combined bending and torsion, there is a truss-like behaviour up to failure as in pure torsion. If bending dominates, the behaviour is similar to that of pure bending except that the compression zone becomes inclined. The two different approaches discussed previously, namely the space truss theory and the skew bending theory, have been extended to explain the interaction behaviour.

Assuming a parabolic interaction behaviour between the pure bending capacity M_u, given by the conventional bending formula, and the pure torsional capacity T_u, given by the space truss theory, the following equations were obtained by Lampert and Thurlimann (1971) (see Fig. 5.5).

(a) For the yielding of the 'bottom' (flexural tensile zone) longitudinal and transverse steel:

$$m\left[\frac{T_{uc}}{T_u}\right]^2 + \left(\frac{M_{uc}}{M_u}\right)^2 = 1 \tag{5.36}$$

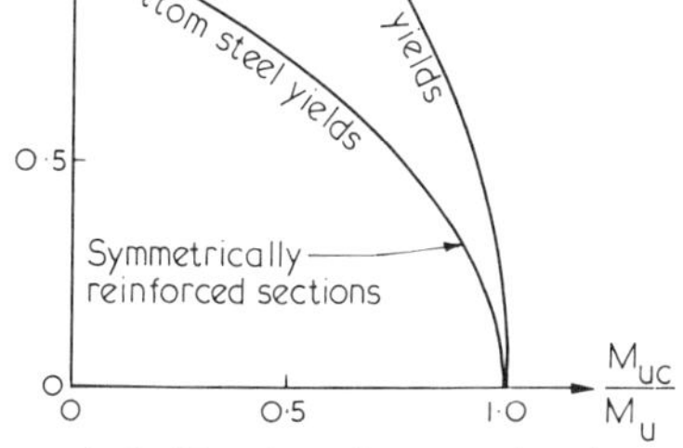

FIG. 5.5. Lampert and Collins bending–torsion interaction diagram.

(b) For the yielding of the top longitudinal and transverse reinforcement:

$$\left(\frac{T_{uc}}{T_u}\right)^2 - \frac{1}{m}\left(\frac{M_{uc}}{M_u}\right)^2 = 1 \qquad (5.37)$$

where M_{uc} and T_{uc} are the ultimate bending and torsional capacity respectively for combined loading.

Kuyt (1972) developed the following interaction equations, which were also based on space truss theory.

(a) For symmetrical arrangement of the longitudinal bars:

$$\left(\frac{T_{uc}}{T_u}\right)^2 + \frac{M_{uc}}{M_u} = 1 \qquad (5.38)$$

(b) For unsymmetrical arrangement of the longitudinal bars:

$$\left(\frac{T_{uc}}{T_u}\right)^2 + \frac{M_{uc}}{M_u} = \alpha' \qquad (5.39)$$

where $\tan^2 \alpha' = \dfrac{\rho_l f_{ly}}{\rho_s f_{sy}}$

$$M_u = \tfrac{1}{2} A_l f_{ly} h \qquad (5.40)$$

T_u is calculated from eqns (5.12) and (5.13) and h is the internal lever arm for bending.

Rabbat and Collins (1978) presented a variable angle space truss model capable of predicting the post-cracking response of reinforced and prestressed concrete members subjected to complex loading. The angles of inclination of the compression diagonal in the walls of the truss model were determined from strain compatibility conditions.

Cowan and Armstrong (1955) have also reported on prestressed beams in combined bending and torsion with various ratios of bending moment to torque, ranging from pure bending to pure torsion. They observed that the web steel reached the ultimate capacity in combined bending and torsion when the first crack formed and the failure was sudden and destructive regardless of the ratio of bending to torque. Vertical bending cracks were observed on the tension side as well as some inclined torsional cracks at about mid-height of the beam. They found that in prestressed concrete beams, as in reinforced concrete beams, the addition of a small amount of bending increased the torsional strength, while a small increase of torsion reduced the bending resistance.

Tests on prestressed concrete hollow and solid beams subjected to combined bending and twisting were also performed on 24 beams by Swamy (1962). He pointed out that, depending on the bending to torque ratio, two different modes of failure were possible: diagonal tension failure and crushing of concrete in compression. He observed that the presence of small amounts of torque did not reduce the load carrying capacity of a beam appreciably; a slight increase in torsional strength was found for bending moment–torque ratios of up to 2·0. He explained that the increase in torsional strength was due to a reduction in tensile stresses, resulting from the effect of the compressive stresses due to bending. For a bending–torque ratio of 3 to 5, a transition from initial cracking due to bending to an ultimate torsional failure was observed. He also stated that no single existing criterion gave consistent results for all stress combinations.

Evans and Khalil (1970) reported tests on 38 centrally and eccentrically prestressed rectangular concrete beams under varying ratios of bending to torsion; 15 of the specimens were eccentrically prestressed and contained web reinforcement. They found that the eccentric prestressing had an advantage in increasing the torsional capacity of a member over pure torsional strength in the presence of flexure so long as the beam did not crack before failure. The failure of conventional prestressed beams as well as those containing web reinforcement were always of the cleavage (tension) type and in a few cases, where the failing bending moment was very high, crushing took place at the top (compression) face of the beam. The use of web reinforcement made failure more progressive and gave ample warning of impending failure. The provision of web reinforcement up to a limiting value was found to increase the torsional capacity of a prestressed concrete member.

Ganga Rao and Zia (1973) tested 36 rectangular beams under combined bending and torsion. The beams were centrally and eccentrically prestressed with the addition of transverse and/or longitudinal reinforcement. They found that the torsional cracking caused a very substantial reduction in rotational stiffness but had a relatively small effect on the bending stiffness. They also reported that with increasing bending moment to torque ratio, the margin between cracking and ultimate strength increased; likewise, the corresponding margin for rotation and deflection also increased. They observed that the failure of the prestressed concrete beams was similar to that of the reinforced concrete beams under combined loading. For low bending to torque ratios, failure was found to take place in the torsion mode of skew bending (Mode II). For moderate to high bending to torque ratios, failure was in the mode of skew bending (Mode I). They suggested a square non-dimensional interaction curve for prestressed beams supplemented with longitudinal steel, and a circular interaction curve for beams without longitudinal steel.

Zia (1970) found that, in general, the interaction between torsion and bending in prestressed concrete members was similar to that of reinforced concrete members. The centrally prestressed beam is analogous to the symmetrically reinforced beam, while the eccentrically prestressed beam is analogous to the unsymmetrically reinforced beam. However, the eccentrically prestressed specimens benefitted from the application of a bending moment up to about 80% of the ultimate bending capacity, after which the strength dropped rapidly with increased moment.

5.6 COMBINED BENDING, TORSION AND SHEAR

In most cases torsion occurs in association with transverse shear and often they are both at peak levels at the same section of the beam.

For a beam loaded with a pure torsional moment T, diagonal tensile stresses will spiral around the beam faces as shown in Fig. 5.6(a). For a beam loaded with flexural shear force V, diagonal tensile stresses will also appear but they will be parallel on the two vertical faces of the beam according to Fig. 5.6(b). Consequently, for a beam loaded in combined torsion and shear the diagonal tensile stresses will be additive on one of the vertical faces and subtractive on the other (see Fig. 5.6(c)). On the horizontal faces the diagonal tensile stresses will be due to torsion only. Limiting the maximum diagonal tensile stress to the tensile strength of the concrete will give a design criterion

$$v_t + v_s \leqslant f_t \qquad (5.41)$$

where v_s is the diagonal tensile stress due to transverse shear. The above criterion leads to a linear interaction diagram between torsion and shear.

In a finite length of a beam, it is impossible to apply a combined loading of torsion and shear without introducing bending; the condition of combined torsion and shear (without bending moment) may be

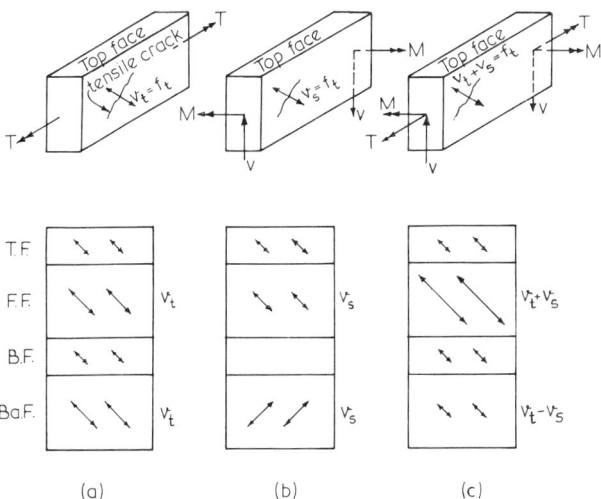

FIG. 5.6. Diagonal tensile stresses due to torsion and flexural shear for rectangular sections. (a) Pure torsion; (b) flexural shear; (c) combined torsion and flexural shear.

achieved only at the points of inflection of continuous beams. This is largely theoretical because in real structures loading is never constant, so that the points of inflection are not stationary; hence the torsion-shear interaction could be considered only as the limiting case of combined torsion, bending and shear. However, Kemp (1968) found that the application of an applied torque would set up a state of shear within the loaded member; hence a greater interaction between flexural shear and torsional shear could be expected than between torque and bending moment.

Nylander (1945) observed that the failure of members without web reinforcement subjected to combined bending, torsion and shear was controlled by the principal tensile stress resulting from the shear stress due to torsion and transverse shear. On this basis he recommended a linear interaction diagram between torsion and shear. Mirza and McCutcheon (1969) showed that the amount of longitudinal reinforcement had a significant influence on the shape of the torsion and shear interaction curves; they proposed a linear interaction diagram (similar to Nylander) which was the lower bound to all other tests. Hsu (1968d) analyzed the available test results of members without web reinforcement and suggested interaction curves, which depend upon the applied moment–flexural capacity of the member as given below:

(a) For $\dfrac{M}{M_u} \leqslant 0.5$,

$$\left(\frac{T}{T_{uc}}\right)^2 + \frac{V}{V_u} = 1 \quad \text{where} \quad T_{uc} = T_u \quad (5.42)$$

(b) For $0.5 \leqslant \dfrac{M}{M_u} \leqslant 1.0$,

$$\left(\frac{T}{T_{uc}}\right)^2 + \left(\frac{V}{V_u}\right)^2 = 1 \quad \text{where} \quad T_{uc} = T_u\left(1.7 - 1.4\frac{M}{M_u}\right) \quad (5.43)$$

In eqns (5.42) and (5.43) V_u is the shear strength based on diagonal cracking in combined shear and bending.

Mattock (1968) proposed a circular interaction curve between shear and torsion similar to eqn (5.43):

$$\left(\frac{v_{tc}}{v_t}\right)^2 + \left(\frac{v_{sc}}{v_s}\right)^2 = 1 \quad (5.44)$$

where v_{tc} is the torsional shear stress carried by concrete when transverse shear and torsion act together, and v_{sc} is the flexural shear stress carried by concrete when transverse shear and torsion act together.

Klus (1968) suggested a bilinear interaction curve between torsion and shear in members containing web reinforcement. However, Mirza's tests (1968) showed that the interaction curves were also affected by the amount of web reinforcement.

Osburn et al. (1969) tested rectangular and T-shaped beams containing web reinforcement subjected to combined torsion, shear and bending, and showed that Mattock's method and those contained in 'Tentative recommendations for the design of reinforced concrete members to resist torsion' by ACI Committee 438 (1969) were somewhat conservative. They also found that the torsional stiffness at service loads of reinforced concrete beams with web reinforcement subjected to combined torsion, shear and bending was approximately half the torsional stiffness of the uncracked section. However, the deflection of such beams when subjected to combined torsion, shear and flexure was not significantly different from that which would occur in similar beams subjected to shear and flexure only.

DESIGN OF CONCRETE STRUCTURES FOR TORSION   161

The studies of McMullen and Warwaruk (1970) discussed previously also included the influence of transverse shear on the torsional strength of reinforced concrete beams with web reinforcement. They found that reinforced concrete beams when subjected to bending, torsion and moderate amounts of transverse shear could fail by the three different modes outlined in Fig. 5.2. These modes of failure were found to be characterized by the formation of a hinge adjacent to one of the faces of the beam and the yielding of reinforcement adjacent to the face opposite the hinge. In beams subjected to a constant twisting moment and a varying bending moment, Mode III failure would occur in the region subjected to the lowest bending moment and Mode I failure would occur in the region subjected to the highest bending moment. Also a small or moderate amount of transverse shear did not significantly affect the torsional strength of a beam failing by Modes I or III.

Analogous to the interaction curves, the effect of combined torsion, shear and bending have been expressed in terms of an interaction surface, determined experimentally or theoretically; such interaction surfaces proposed by various investigators for beams with and without web reinforcement are reviewed by Zia (1970).

Ewida and McMullen (1981) proposed an interaction surface for torsion-shear-flexure, which was actually a combination of two surfaces (see Fig. 5.7). The equation defining torsion-shear interaction is given by

$$\frac{T_{uc}}{r \cdot T_{url}} + \left[\frac{V_{uc}}{V_u}\right]^n = 1 \tag{5.45}$$

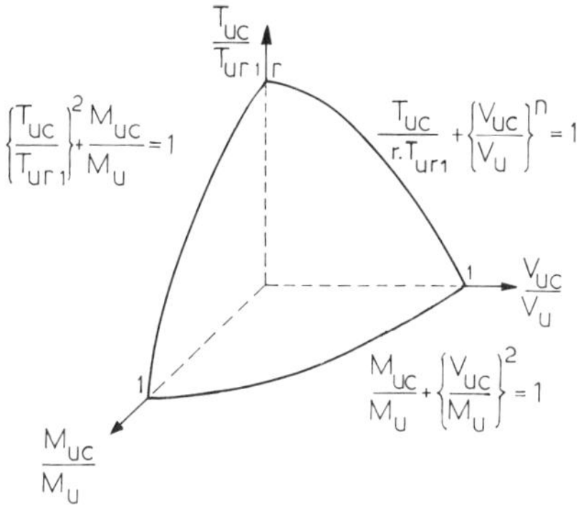

FIG. 5.7. Ewida and McMullen's theoretical torsion-shear-flexure interaction surface.

where $n = 1\cdot2$ for beams under-reinforced in torsion–shear,
$n = 1\cdot75$ for beams partially over-reinforced in torsion–shear,
$n = 3\cdot0$ for beams substantially over-reinforced in torsion–shear.

$$r = \frac{T_{u2}}{T_{u1}}$$

where T_{u1} is the ultimate torque in pure torsion for Mode I and T_{u2} is the ultimate torque in pure torsion for Mode II.

Tests on prestressed concrete rectangular I- and T-sections carried out by Bishara (1969) showed that the simultaneous action of torque, moment and shear reduced the capacities of the tested beams below their corresponding ultimate capacities under the action of moment and shear only. The percentage reduction was found to be related to M/T ratio. The maximum torsional capacity was reached at M/T ratio equal to $0\cdot29\,M_u/T_u$. The shape of interaction curves of his test results showed that only for very high M/T ratios ($M/T > 0\cdot8\,M_u/T_u$) did the torsional capacity drop below that in pure torsion. The following relationships between the moment and torque or torque and shear at ultimate capacity were proposed by him for rectangular sections:

$$\frac{T_{uc}}{T_u} = 1\cdot0 + 2\cdot3\left(\frac{M_{uc}}{M_u}\right) - 3\cdot3\left(\frac{M_{uc}}{M_u}\right)^2 \tag{5.46}$$

$$\frac{T_{uc}}{T_u} = 1\cdot0 + 2\cdot3\left(\frac{V_{uc}}{V_u}\right) - 3\cdot3\left(\frac{V_{uc}}{V_u}\right)^2 \tag{5.47}$$

Gausel (1970) tested prestressed I-beams reinforced with and without web reinforcement subjected to combined torsion, bending and shear. He found that the web reinforcement had a small effect on ultimate strength in torsion, but a very small amount of web reinforcement would prevent explosive failures, while at the same time the twisting capacity would increase. His tests indicated that the ultimate loads were independent of the loading sequence of torsion and bending/shear.

The studies of Mukherjee and Warwaruk (1971) (reviewed previously) also included prestressed concrete rectangular members containing non-prestressed longitudinal and transverse reinforcement subjected to combined bending, shear and torsion. They found that the initial torsional stiffness was practically unaffected by the level of prestress, torque–moment ratio or flexural shear; the point of departure and the rate of deviation from the initial stiffness was dependent on the torque–

moment ratio. They also found that for any combination of torsion and bending, the presence of flexural shear was detrimental to the torsional strength.

Rangan and Hall (1973) developed strength equations for rectangular prestressed concrete beams subjected to combined bending, shear and torsion along the lines of previously developed equilibrium methods for reinforced concrete beams. From these they derived three equations to predict the strengths of beams failing in Modes I, II and III. For the prediction of strength in the case of shear-compression mode of failure, they proposed a fourth empirical equation.

Henry and Zia (1974) tested centrally and eccentrically prestressed rectangular concrete beams containing longitudinal steel and web reinforcement subjected to combined bending, shear and torque. They found that the beams tended to fail in one of two modes of skewed bending, viz. torsion mode or bending mode. The bending mode of failure occurred for values of T/M approximately equal to or less than 0·22; for larger values, the torsion mode occurred. They also found that when the beam was under the action of combined torsion, bending and shear, the ultimate torsional capacity of the beam could be increased up to approximately 30% when the bending moment and shear were kept low (i.e. below 25% of the ultimate flexural capacity). However, the presence of shear in combined loads reduced the ultimate torsional capacity of a beam by 10–20%, as compared with beams loaded in torsion and bending only.

5.7 RECENT DEVELOPMENTS AND DESIGN RECOMMENDATIONS

According to Hsu (1984) recent developments in space truss theories have taken place by developing theoretical models based on the following concepts:

(a) the *compression field theory* based on a variable angle truss model, where the angle of inclination of the cracks is assumed to be the same as the angle of inclination of the compression field;
(b) the *plasticity compression field theory* based on the variable angle truss model due to Lampert and Thurlimann (1971), and extended to members subjected to torsion, bending and shear by Elfgren (1972), using the plastic theory;

(c) the *compatibility compression field theory* wherein the strain compatibility of the truss model has been accounted for.

The CEB-FIP Model Design Code (1978) is based on the plasticity compression field theory; this code separates torsion into two types, viz. equilibrium torsion and compatibility torsion. When a torsional moment is required to be applied to ensure the equilibrium of a structure, it is termed 'equilibrium torsion'; on the other hand, torque induced in order to maintain compatibility of deformation is termed 'compatibility torsion'. The former cannot be reduced by the redistribution of internal forces within the structure, while the latter can. The code requires that full torsional moment must be accounted for when equilibrium torsion is encountered; however, compatibility torsion may be neglected when considering the ultimate capacity of the member.

Collins and Mitchell (1980) have proposed design recommendations using the compatibility compression field theory, which enable prestressed and non-prestressed concrete beams to be designed for shear and torsion. A range of allowable angles of diagonal compression are suggested, based on the considerations of concrete crushing and diagonal cracking. A merit of this method is that the same basic expressions are employed for the design of non-prestressed, partially prestressed and fully prestressed beams having a wide variety of cross sections.

Recently, Hsu and Mo (1985) have pointed out that the previous investigators had made arbitrary (and theoretically unjustified) modifications to Rausch's equation (eqn 5.10) in order to interpret the observed test results. These modifications may be classified into three categories:

(a) by the addition of an efficiency factor for reinforcement (see Anderson, 1935; Hsu, 1968*c*)
(b) by an arbitrary definition for the centre line of shear flow (see Lampert and Thurlimann, 1968; CEB-FIP, 1978);
(c) by the deletion of concrete cover (see Collins and Mitchell, 1980).

Hsu and Mo have shown that the deviation between the theoretical predictions and experimentally observed values can be explained by the softening of concrete due to diagonal cracking. Using a new stress-strain relationship for concrete which accounts for this softening they have proposed a modified space truss theory which is validated by the test results. A complete set of design recommendations have been proposed using this modified theory which accounts for design limitations of over-reinforcement, minimum reinforcement and concrete cover.

REFERENCES

ACI COMMITTEE 438 (1969) Tentative recommendations for the design of reinforced concrete members to resist torsion. *Journal of the American Concrete Institute,* **66** (1), 1–8.

ACI STANDARD 318-71 (1971) *Building Code Requirements for Reinforced Concrete,* American Concrete Institute, Detroit, 78 pp.

ANDERSON, P. (1935) Experiments with concrete in torsion. *Transactions, ASCE,* **100**, paper no. 1912, pp. 949–83.

ANDERSON, P. (1937) Rectangular concrete sections under torsion. *Journal of the American Concrete Institute,* Proceedings **34**(1), 1–11.

BACH, B. and GRAF, O. (1912) Versuche uber Widerstands Fahigkeit von Beton und Eisenbeton gegen Verdrehung, *Deutscher Ausschuss für Eisenbeton,* Heft 16, Wilhelm Ernst, Berlin.

BISHARA, A. (1969) Prestressed concrete beams under combined torsion, bending and shear. *Journal of the American Concrete Institute,* **66** (7), 525–38.

BISHARA, A. and PEIR, J. C. (1968) Reinforced concrete rectangular columns in torsion. *Journal of the Structural Division, ASCE,* **94** (ST12), 2913–33.

CEB–FIP (1978) *Model Code for Concrete Structures,* CEB–FIP International Recommendations, Comité Euro-International du Béton, Paris.

CHAKRABORTY, M. (1979) Ultimate torque of reinforced rectangular beams. *Journal of the Structural Division, ASCE,* **105** (ST3), 651–68.

COLLINS, M. P. and MITCHELL, D. (1980) Shear and torsion design of prestressed and nonprestressed concrete beams. *Journal of the Prestressed Concrete Institute,* **25** (5), 32–100.

COLLINS, M. P., WALSH, P. F., ARCHER, F. E. and HALL, A. S. (1968) Ultimate strength of reinforced concrete beams subjected to combined torsion and bending. In *Torsion of Structural Concrete,* SP18, American Concrete Institute, Detroit, pp. 379–402.

COWAN, H. J. (1951) Tests of torsional strength and deformation of rectangular reinforced concrete beams. *Concrete and Constructional Engineering, London,* **46**(2), 51–9.

COWAN, H. J. (1953) The strength of plain, reinforced and prestressed concrete under the action of combined stresses, with particular reference to the combined bending and torsion of rectangular sections. *Magazine of Concrete Research, London,* **5** (14), 75–86.

COWAN, H. J. and ARMSTRONG, S. (1955) Experiments on the strength of reinforced and prestressed concrete beams and of concrete-encased steel joists in combined bending and torsion. *Magazine of Concrete Research, London,* **7** (19), 3–20.

ELFGREN, L. (1972) *Reinforced Concrete Beams Loaded in Combined Bending, Torsion and Shear,* Publication 71-3, Division of Concrete Structures, Chalmers University of Technology, Gotenburg, Sweden.

EVANS, R. H. and KHALIL, M. G. (1970) The behaviour and strength of prestressed concrete and rectangular beams subjected to combined bending and torsion. *Structural Engineer, London,* **48** (2), 59-73.

EWIDA, A. A. and MCMULLEN, A. E. (1981) Torsion-shear-flexure interaction in reinforced concrete members. *Magazine of Concrete Research, London,* **33** (115), 113-22.

GANGA RAO, H. V. S. and ZIA, P. (1973) Rectangular prestressed beams in torsion and bending. *Journal of the Structural Division, ASCE,* **99** (ST1), 183-98.

GAUSEL, E. (1970) Ultimate strength of prestressed I-beams under combined torsion, bending and shear. *Journal of the American Concrete Institute,* **67** (9), 675-8.

GOODE, C. D. and HELMY, M. A. (1968) Ultimate strength of reinforced concrete beams in combined bending and torsion. In *Torsion of Structural Concrete,* SP18, American Concrete Institute, Detroit, pp. 357-78.

HENRY, R. L. and ZIA, P. (1974) Prestressed beams in torsion, bending and shear. *Journal of the Structural Division, ASCE,* **100** (ST5), 933-52.

HSU, T. T. C. (1968a) Torsion of structural concrete — plain concrete rectangular sections. In *Torsion of Structural Concrete,* SP18, American Concrete Institute, Detroit, pp. 203-38.

HSU, T. T. C. (1968b) Ultimate torque of reinforced rectangular beams. *Journal of the Structural Division, ASCE,* **94** (ST2), 485-510.

HSU, T. T. C. (1968c) Torsion of structural concrete — behaviour of reinforced concrete rectangular members. In *Torsion of Structural Concrete,* SP18, American Concrete Institute, Detroit, pp. 261-306.

HSU, T. T. C. (1968d) Torsion of structural concrete — interaction surface for combined torsion, shear, and bending in beams without stirrups. *Journal of the American Concrete Institute,* **65** (1), 51-60.

HSU, T. T. C. (1984) *Torsion of Reinforced Concrete,* Van Nostrand Reinhold Company, New York.

HSU, T. T. C. and KEMP, E. L. (1969) Background and practical application of tentative design criteria for torsion. *Journal of the American Concrete Institute,* **66** (1), 12-23.

HSU, T. T. C. and MO, Y. L. (1985) Softening of concrete in torsional members: design recommendations. *Journal of the American Concrete Institute,* **82**(4), 443-52.

HUMPHREYS, R. (1957) Torsional properties of prestressed concrete. *Structural Engineer, London,* **35** (6), 213-24.

IYENGAR, K. T. S., DESAYI, P. and SETHUNARAYANAN, S. (1974) Discussion of tests of concrete columns in torsion. *Journal of the Structural Division, ASCE,* **100** (ST5), 1150-3.

KEMP, E. L. (1968) Behaviour of concrete members subject to torsion and to combined torsion, bending and shear. In *Torsion of Structural Concrete,* SP18, American Concrete Institute, Detroit, pp. 179-201.

KLUS, J. P. (1968) Ultimate strength of reinforced concrete beams in combined torsion and shear. *Journal of the American Concrete Institute,* **65** (3), 210-15.

KUYT, B. (1971) A theoretical investigation of ultimate torque as calculated by truss theory and by the Russian ultimate equilibrium method. *Magazine of*

Concrete Research, London, **23** (77), 155-60.
KUYT, B. (1972) A method of ultimate strength design of rectangular reinforced concrete beams in combined torsion, bending and shear. *Magazine of Concrete Research, London,* **24** (78), 15-24.
LAMPERT, P. and COLLINS, M. P. (1972) Torsion, bending and confusion — an attempt to establish facts. *Journal of the American Concrete Institute,* **69** (8), 500-4.
LAMPERT, P. and THURLIMANN, B. (1968) *Torsion Tests on Reinforced Concrete,* Report No. 6502-2, Institute of Structural Engineering, Swiss Federal Institute of Technology, Zurich.
LAMPERT, P. and THURLIMANN, B. (1971) Ultimate strength and design of reinforced concrete beams in torsion and bending. *Publications, International Association for Bridge and Structural Engineering (Zurich),* Vol. 31-I, pp. 107-31.
LESSIG, N. N. (1959) Determination of the load-bearing capacity of reinforced concrete elements with rectangular cross section subjected to flexure with torsion, *Trudy,* No. 5, Concrete and Reinforced Concrete Institute, Moscow, pp. 5-28. (Translated as PCA Foreign Literature Study No. 371.)
MARSHALL, W. T. (1974) Torsion in concrete and CP110. *Structural Engineer, London,* **52** (3), 83-8.
MATTOCK, A. H. (1968) How to design for torsion. In *Torsion of Structural Concrete,* SP18, American Concrete Institute, Detroit, pp. 469-91.
MCMULLEN, A. E. and WARWARUK, J. (1970) Concrete beams in bending, torsion and shear. *Journal of the Structural Division, ASCE,* **96** (ST5), 885-903.
MIRZA, M. S. (1968) Discussion of ultimate strength of reinforced concrete beams in combined torsion and shear. *Journal of the American Concrete Institute,* **65** (9), 788-90.
MIRZA, M. S. and MCCUTCHEON, J. O. (1969) Behaviour of reinforced concrete beams under combined bending, shear and torsion. *Journal of the American Concrete Institute,* **66** (5), 421-7.
MITCHELL, D. and COLLINS, M. P. (1974) Diagonal compression field theory — a rational model for structural concrete in pure torsion. *Journal of the American Concrete Institute,* **71** (8), 396-408.
MUKHERJEE, P. R. and WARWARUK, J. (1971) Torsion bending and shear in prestressed concrete. *Journal of the Structural Division, ASCE,* **97** (ST4), 1063-79.
NADAI, A. (1950) *Theory of Flow and Fracture of Solids,* McGraw Hill Book Company, New York.
NARAYANAN, R. and KAREEM-PALANJIAN, A. S. (1983) Steel fibre reinforced concrete beams in torsion. *International Journal of Cement Composites and Lightweight Concrete,* **5** (4), 235-46.
NYLANDER, H. (1945) *Torsion and Torsional Resistance of Concrete Structures,* Statens Kommittee for Bygnadsforskning, Stockholm, Meddlelanden 3 (no. 3).
OSBURN, D. L., MAYOGLOU, B. and MATTOCK, A. H. (1969) Strength of reinforced concrete beams with web reinforcement in combined torsion, shear and bending. *Journal of the American Concrete Institute,* **66** (1), 31-41.

PANDIT, G. S. (1970) Ultimate torque of rectangular reinforced concrete beams. *Journal of the Structural Division, ASCE,* **96** (ST9), 1987-95.

PANDIT, G. S. and MAWAL, B. M. (1973) Tests of concrete columns in torsion. *Journal of the Structural Division, ASCE,* **99** (ST7), 1409-21.

PANDIT, G. S. and VENKAPPA, V. (1980) Ultimate torque expression — a critical reappraisal. *Indian Concrete Journal,* **54** (12), 320-5.

RABBAT, B. G. and COLLINS, M. P. (1978) A variable angle space truss model for structural concrete members subjected to complex loading. In *Douglas McHenry International Symposium on Concrete and Concrete Structures,* SP55, American Concrete Institute, Detroit, pp. 547-87.

RAJAGOPALAN, K. S., BEHERA, U. and FERGUSON, P. M. (1971) Partially over-reinforced concrete beams under pure torsion. *Journal of the American Concrete Institute,* **68** (10), 740-7.

RANGAN, B. V. and HALL, A. S. (1973) Strength of rectangular prestressed concrete beams in combined torsion, bending and shear. *Journal of the American Concrete Institute,* **70** (3), 270-8.

RAUSCH, E. (1929) *Berechnung des Eisenbetons gegen Verdrehung und Abscheven* (Design of reinforced concrete for torsion and shear), Springer-Verlag, Berlin, p. 50.

SAINT-VENANT, B. DE (1856) *Mémoire sur la Torsion des Prismes,* Mémoires présentés par divers savants a l'Académie des Sciences, de l'Institut Impérial de France et Imprimé par son ordre, Vol. 14, Imprimerie Impériale, Paris, 1856.

SWAMY, N. (1962) The behaviour and ultimate strength of prestressed concrete hollow beams under combined bending and torsion. *Magazine of Concrete Research, London,* **14** (40), 13-24.

VICTOR, D. J. and MUTHUKRISHNAN, R. (1973) Effect of stirrups on ultimate torque of reinforced concrete beams. *Journal of the American Concrete Institute,* **70** (4), 300-6.

WARNER, R. F., RANGAN, B. V. and HALL, A. S. (1977) *Reinforced Concrete,* Pitman Publishing Ltd, Australia, p. 134.

YUDIN, V. K. (1962) Determination of the load-bearing capacity of reinforced concrete elements of rectangular cross section. *Trudy,* No. 6, Concrete and Reinforced Concrete Institute, Moscow, pp. 265-8. (Translated as PCA Foreign Literature Study No. 377.)

ZIA, P. (1961) Torsional strength of prestressed concrete members. *Journal of the American Concrete Institute,* **57** (10), 1337-59.

ZIA, P. (1968) Torsion theories for concrete members. In *Torsion of Structural Concrete,* SP18, American Concrete Institute, Detroit, pp. 103-32.

ZIA, P. (1970) What do we know about torsion in concrete members? *Journal of the Structural Division, ASCE,* **96** (ST6), 1185-99.

ZIA, P. and MCGEE, W. D. (1974) Torsion design of prestressed concrete. *Journal of the Prestressed Concrete Institute,* **19** (2), 46-65.

Chapter 6

REINFORCED CONCRETE DEEP BEAMS

F. K. KONG

Department of Civil Engineering, University of Newcastle upon Tyne, UK

SUMMARY

This chapter reviews the behaviour, analysis and design of reinforced concrete deep beams with and without web openings. The design of deep beams is discussed in relation to the current Codes of Practice. Some of the main recommendations of the comprehensive deep-beam guide, issued by the Construction Industry Research and Information Association in 1977, are summarized and a design example is given.

NOTATION

A	Area of a reinforcement bar (see eqn 6.1)
A_s	Area of main longitudinal reinforcement
a_v	Shear span (see Fig. 6.1)
b	Beam thickness
C, C_1, C_2	Support lengths (see eqn 6.2)
d	Effective depth
$f_{cu}(f_y)$	Characteristic strength of concrete (of reinforcement)
h	Overall height of beam
h_a	Effective height of beam (see eqn 6.3)
k_1, k_2	Coefficients defining the position of an opening (see Fig. 6.8)
l	Effective span (see eqn 6.2)
l_0	Clear distance between faces of supports
M	Ultimate moment (see eqns 6.4 and 6.5)

V	Ultimate shear force; shear capacity
v_c	Shear stress value (see eqn 6.7)
v_u	Ultimate shear stress value (see eqn 6.8)
$v_x, v_{ms}, v_{wh}, v_{wv}$	Shear stress parameters (see eqn 6.11)
x	Clear shear span (see eqn 6.1)
x_e	Effective clear shear span (see eqn 6.10)
y, y_1	Distances defined in eqns (6.1) and (6.13), respectively
z	Lever arm
$\theta, \theta_r, \theta_1$	Angles defined in eqns (6.1), (6.10) and (6.13) respectively
ρ	Steel ratio
ϕ	Angle (see Figs 6.3 and 6.9)

6.1 INTRODUCTION

A beam having a depth comparable to the span length is called a deep beam. The design of reinforced concrete deep beams is a topic which recurs in practice, particularly in the design of tall buildings, offshore structures and foundations. However, it is only comparatively recently that research on reinforced concrete deep beams was carried out on a practical scale (Albritton, 1965; C & CA, 1969), and even more recently that design recommendations were given in official documents. In 1970 the Comité Européan du Béton and the Fédération Internationale de la Précontrainte first included provisions for deep beams in their International Recommendations (CEB–FIP, 1970). In 1971, the American Code for the first time included recommendations for deep beams (ACI 318-71, 1971). In 1977, recognizing industry's need for data and guidance, the Construction Industry Research and Information Association in the UK issued CIRIA Guide 2: *The Design of Deep Beams in Reinforced Concrete;* this 131 page design guide, prepared by Ove Arup and Partners, became the most comprehensive document of its kind in the English speaking countries. Currently, the three major documents covering deep-beam design are ACI 318-83 (1983), the CEB–FIP Model Code (1978) and CIRIA Guide 2 (1977). However, the deep-beam provisions in ACI 318-83 are essentially the same as those in the 1971 code; similarly the deep-beam provisions in the CEB–FIP Model Code (1978) remain essentially the same as those in the 1970 CEB–FIP International Recommendations. Hence, to this day, the CIRIA Guide remains not only the most comprehensive deep-beam guide available, but it is also in fact the most up to date one.

REINFORCED CONCRETE DEEP BEAMS 171

This chapter explains the behaviour of reinforced concrete deep beams and their design to comply with the respective requirements of ACI 318-83, the CEB–FIP Model Code and the CIRIA Guide. Recent developments on web openings and on stability and buckling are briefly reviewed.

6.2 DEEP-BEAM BEHAVIOUR

6.2.1 Elastic Behaviour
Under elastic condition, the transition from ordinary-beam behaviour to deep-beam behaviour occurs at a span/depth ratio of about 2·5. A substantial library of work is available on the elastic behaviour of deep beams (Albritton, 1965; C & CA, 1969). Special areas of elastic behaviour not covered by the literature can usually be investigated with relative ease, because the computer has now made possible the ready applicability of finite difference and finite element techniques (Robins and Kong, 1973; Zienkiewicz, 1977; Coates *et al.*, 1980; Cope *et al.*, 1982).

CIRIA (1977) has reviewed the principal publications on the subject and summarized the elastic behaviour of reinforced concrete deep beams. Briefly it can be stated that

(a) plane sections across the beam do not remain plane, and there may be more than one neutral axis;
(b) the distribution of stresses in top-loaded deep beams is noticeably different from that in bottom-loaded deep beams;
(c) the effective height of a deep beam is roughly equal to the span. That part of the beam above this height acts essentially as a load-bearing wall and plays no part in carrying the load between supports;
(d) there is an area of high, biaxial stress over the supports and a tendency for splitting forces to arise under concentrated loads and above supports.

The main disadvantage of elastic studies is that they assume an isotropic material obeying Hooke's law, and hence provide inadequate guidance on ultimate load behaviour.

6.2.2 Ultimate Load Behaviour
The ultimate load tests of de Paiva and Siess (1965) and Leonhardt and Walther (1966) marked a significant stage in the move from elastic to ultimate load investigations, which have since been taken up by others

in different parts of the world (see CIRIA, 1977). In the late 1960s an extensive long-term programme was initiated in the UK which is still continuing at the University of Newcastle upon Tyne today (Kong *et al.*, 1986). The research of the past 20 years has contributed much to our understanding of the principal failure modes of reinforced concrete deep beams — shear, bearing, and flexure.

Before discussing the failure modes of deep beams, it is instructive to re-examine the effect of the shear span/depth ratio (a_v/d in Fig. 6.1) on the failure of ordinary reinforced concrete beams, as described in the literature (e.g. Kong and Evans, 1980).

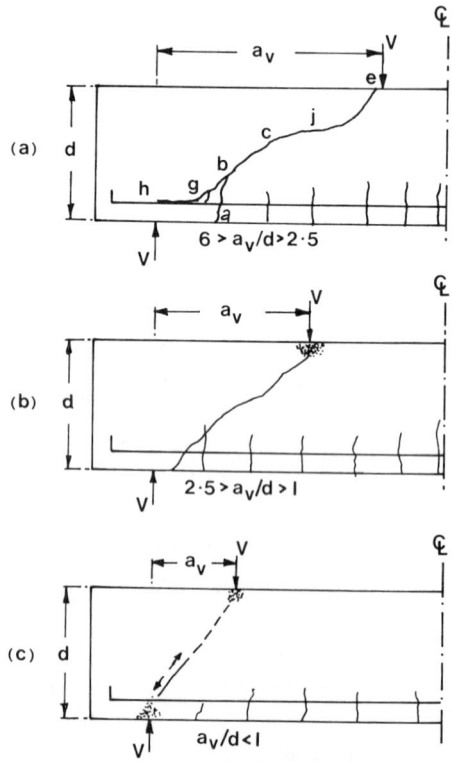

FIG. 6.1. Effect of shear span/depth ratio, a_v/d, on failure mode.

(a) $a_v/d > 6$: Beams with such a high a_v/d ratio usually fail in flexure.
(b) $6 > a_v/d > 2.5$ (Fig. 6.1a): Beams with a_v/d lower than 6 tend to fail in shear. With reference to Fig. 6.1(a), as the shear force V is increased, the flexural crack a–b nearest the support would propagate towards the loading point, gradually becoming a diagonal crack (a–b–c in Fig. 6.1a). If the a_v/d ratio is relatively high, the diagonal crack would rapidly spread to point e, splitting the beam into two pieces. If the a_v/d ratio is relatively low, the diagonal crack tends to stop at j (Fig. 6.1a). As V is further increased, failure occurs through the propagation of the diagonal crack along the level of the main reinforcement (Fig. 6.1(a): crack g–h). For either failure mode, the ultimate load is not much higher than the diagonal-cracking load, i.e. that at which the crack b–c forms.
(c) $2.5 > a_v/d > 1$ (Fig. 6.1b): Here the diagonal crack forms independently and not as a development of a flexural crack. The beam usually remains stable after such cracking. With further increase in the shear force V, the diagonal crack penetrates into the concrete compression zone at the loading point; eventually crushing failure of the concrete occurs there. For this failure mode, the ultimate load may be more than twice the diagonal-cracking load.
(d) $a_v/d < 1$ (Fig. 6.1c): Here, the diagonal crack forms approximately along a line joining the loading and support points. It frequently initiates at about $d/3$ above the bottom face of the beam, and propagates simultaneously towards the loading and support points, causing crushing failure of the concrete. The ultimate load is often several times the diagonal-cracking load.

Of the three types of diagonal cracks identified above, the type most relevant to deep beams is that in Fig. 6.1(c). Typically, when a load is applied to a deep beam, the first cracks to form are flexural cracks within the central half of the beam (Kong and Singh, 1972). These vertical cracks propagate rather rapidly after formation. As the load is increased, further cracks will form near the supports and propagate towards the loading points. These cracks, which have the common property of initiating at or very near the beam, tend to be quite harmless unless the main tension reinforcement is particularly light. At higher loads, say 50–90% of the ultimate, a new type of diagonal cracks would suddenly appear within the shear span, often accompanied by a loud noise. These diagonal cracks, whose formation is akin to the splitting of the concrete in a standard split-cylinder test, are the dangerous cracks and they have the common property of initiating not at the soffit but at about a third of the beam depth above the soffit. They are often quite long even when they first appear. They propagate simultaneously towards the loading point and the support, in the general direction of the line joining the loading and support points. Kong and Singh (1972) and Kong and Sharp (1973, 1977) have used the term critical diagonal crack to refer to a diagonal crack which subsequently has a direct effect on the failure of the beam. Depending on the effectiveness of the web reinforcement, the critical diagonal crack(s) may cause collapse in one of three principal failure modes:

Mode A (Fig. 6.2a): If the web reinforcement is not effective, the critical diagonal crack may split the beam from top to bottom, without crushing of the concrete.

Mode B (Fig. 6.2b): Where the web reinforcement is comparatively more effective, the beam fails by the crushing of the strut-like portion of concrete between two diagonal cracks.

Mode C (Fig. 6.2c): With a further increase in the effectiveness of the web reinforcement, collapse is delayed until the critical diagonal crack penetrates into the concrete compression zone at the loading or support region; the beam then fails by the crushing of the concrete there.

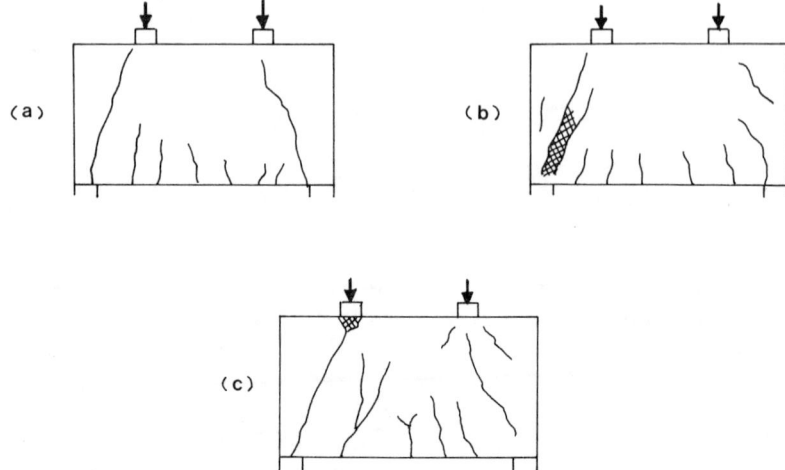

FIG. 6.2. Principal modes of failure. (After CIRIA Guide, 1977.)

If the web reinforcement can arrest the growth of the diagonal cracks and prevent their penetration into the loading or support region, eventual collapse might occur as a result of the bearing failure of the concrete at the loading or support point. If special reinforcement is provided to confine the concrete at the load-bearing regions, or if loads are applied through sufficiently large load-bearing blocks, then the beam will fail in flexure. Of course, flexural failure will also occur if the main longitudinal reinforcement is very light, particularly where the shear span/depth ratio is comparatively large (de Paiva and Siess, 1965; Leonhardt and Walther, 1966).

6.3 SHEAR STRENGTH

The failure modes in Fig. 6.2 suggest the simple structural idealization in Fig. 6.3, used by Kong *et al.* (1972) and Kong and Sharp (1973) to obtain the shear strength equation (eqn 6.1) below, which was subsequently adopted in Section 3.4 of the CIRIA Guide (1977). The structural idealization in Fig. 6.3 shows the applied load as transmitted to the support along the direct path ABC. The effectiveness of the load path can be expected to increase with the angle ϕ; in other words, the load-carrying capacity of the beam increases as the shear span/depth ratio decreases. When the load V (Fig. 6.3) is sufficiently high, a diagonal crack forms along AC, in a manner similar to the splitting of the concrete in a standard split-cylinder test. After diagonal cracking, the ability of the beam to sustain the load depends on the ability of the web reinforcement to arrest the growth of the diagonal crack. With reference to Fig. 6.3, the effectiveness of a web bar depends not only on its size but also on the angle θ at which it intersects the diagonal crack.

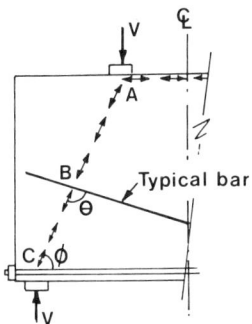

FIG. 6.3. Structural idealization.

Tests to destruction were carried out on about 200 beams (Kong et al., 1972; Kong and Singh, 1974), and the results have shown that where stiff load-bearing blocks are used, the angle θ should be measured from the line joining the inside edge of the bearing block at the support to the outer edge of that at the loading point (Fig. 6.4). The following equation can then be used to calculate the shear resistance of deep beams (Kong et al., 1975):

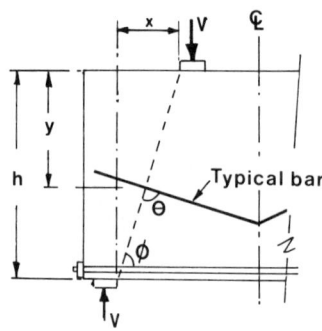

FIG. 6.4. Meanings of symbols in eqn (6.1).

$$V = C_1\left[1 - 0.35\frac{x}{h}\right] f_t bh + C_2 \sum^n A\frac{y}{h}\sin^2\theta \quad (6.1)$$

$$\left(\begin{array}{c}\text{Shear}\\\text{strength}\end{array} = \text{concrete contribution} + \text{steel contribution}\right)$$

where C_1 is a coefficient equal to 1·4 for normal weight concrete and 1·0 for lightweight concrete; C_2 is a coefficient equal to 300 N/mm² for deformed bars and 130 N/mm² for plain round bars; x and h are as explained in Fig. 6.4. For a uniformly distributed load, the x/h ratio is taken as $l/4h$ where l is the span length (Kong et al., 1972); f_t is the cylinder-splitting tensile strength of concrete; if f_t is not available, it may be estimated from the cube strength f_{cu} by, say, $f_t = 0.4$ to $0.5\sqrt{f_{cu}}$; A is the area of a typical web bar — for the purpose of eqn (6.1), the main longitudinal bars are considered also to be web bars (see comment (b) below); y is the depth at which the typical bar intersects the critical diagonal crack, which is represented by the dotted line in Fig. 6.4 (see comment (b) below); θ is the angle between the bar being considered and the diagonal crack (Fig. 6.4); n is the total number of web bars, including the main longitudinal bars (see definition of A above), that intersect the critical diagonal crack.

Comments on eqn (6.1)

(a) Equation (6.1) states that the shear strength of a deep beam is made up of the concrete resistance and the steel resistance. The quantity $C_1 f_t b h$ is a measure of the load-carrying capacity of the 'concrete strut' between the loading and support points. The factor $(1 - 0.35x/h)$ allows for the experimental observation of the way in which this capacity varies with the x/h ratio. The first term in eqn (6.1) is therefore a semi-empirical expression for the capacity of the concrete; when this capacity is reached the 'concrete strut' can be considered to fail in a splitting mode (hence the splitting strength f_t is used), resulting in the formation of the so-called critical diagonal crack.

(b) The second term on the right-hand side of eqn (6.1) represents the contribution of the reinforcement to the load-carrying capacity of the beam; the reinforcement restrains the propagation and widening of the critical diagonal crack, and is compatible with a failure mechanism in which the end portion of the beam moves outwards in a predominantly rotational motion about the loading point (Kong and Sharp, 1973). It can be seen that the ability of a reinforcing bar to restrain such rotation increases with the depth y at which it intersects the critical diagonal crack. This explains why, in eqn (6.1), the steel contribution $C_2 \Sigma A(y/h) \sin^2 \theta$ is proportional to y. The laws of equilibrium are unaware of the designer's discrimination between bars labelled as 'web reinforcement' and bars labelled as 'main reinforcement'. Hence eqn (6.1) accepts any reinforcement bar that participates in the preservation of the integrity of the concrete web, through helping to restrain the growth of the diagonal cracks. It judges the contribution of an individual bar by its area A, the intercepting depth y and the intersecting angle θ.

(c) Equation (6.1) focuses attention on the basic features of what in reality is a complex load-transfer mechanism; it does this by deleting quantities which are less important compared with the main elements — quantities whose inclusion will obscure the designer's understanding of the problem at the physical level. After hundreds of tests (Kong and co-workers, 1972–1985; Besser and Cusens, 1984) it can be said that eqn (6.1) is a useful tool in the hands of engineers who possess a sound understanding of statics, geometry and structural behaviour. Of course, the equation can be abused by indiscriminate applications — as indeed can Codes of Practice be so abused. Consider, for example, a deep beam with a wide bottom flange, which contains two large-diameter longitudinal bars away from the plane of the web. These bars clearly do not effectively protect the integrity of the concrete web, though they have a large product $Ay\sin^2\theta$; hence it would be inappropriate to include such bars when using eqn (6.1).

6.4 DEEP-BEAM DESIGN: ACI 318-83 (1983)

Section 11.8 of the 1983 ACI Code gives special provisions for deep beams. However, these provisions are essentially the same as those in the 1977 and 1971 Codes. Hence the design example and comments given earlier by Kong *et al.* (1975) are still applicable.

6.5 DEEP-BEAM DESIGN: CEB–FIP MODEL CODE (1978)

Section 18.1.8 of the 1978 CEB–FIP Model Code gives special provisions for deep beams. However, these provisions are essentially the same as those in the CEB–FIP International Recommendations (1970). Hence the design example and comments given earlier by Kong *et al.* (1975) are still applicable.

6.6 DEEP-BEAM DESIGN: CIRIA GUIDE (1977)

The CIRIA Guide (CIRIA, 1977) applies to deep beams of rectangular section with an effective span/depth ratio l/h of less than 2 for single-span beams and less than 2·5 for multi-span beams. It is to be used in conjunction with the 1972 British Code (CP110, 1972). The Guide considers that the active height of a deep beam is limited to a depth equal to the span; the portion of the beam above this height is conservatively regarded merely as a load-bearing wall between supports (Besser and Cusens, 1984). With reference to Fig. 6.5, the effective length l and the active height h_a are calculated as follows:

$$l = l_0 + \frac{C_1}{2} + \frac{C_2}{2} \qquad (6.2)$$

where l_0 is the clear span, and the support width C_1 (or C_2) is to be taken as $0.2\, l_0$ if it exceeds $0.2\, l_0$.

$$h_a = h \quad \text{or} \quad l \text{ whichever is the lesser} \tag{6.3}$$

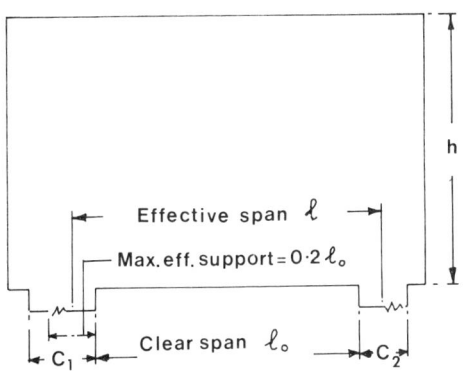

FIG. 6.5. Dimensions of deep beam.

6.6.1 CIRIA's 'Simple Rules'
The CIRIA Guide provides a set of 'simple rules' for uniformly loaded deep beams. These rules cover bending, shear and bearing; their application to single-span beams may be summarized as follows.

6.6.1.1 Strength in Bending
Step 1: If $l/h_a > 1.5$ check that the applied moment M does not exceed the capacity of the concrete section:

$$M < 0.12 f_{cu} b h_a^2 \tag{6.4}$$

where b is the beam thickness.

Step 2: Calculate area A_s of the main longitudinal reinforcement:

$$A_s > \frac{M}{0.87 f_y z} \tag{6.5}$$

where the lever arm z is to be taken as $0.2l + 0.4 h_a$ for a single-span beam.

Step 3: Distribute the reinforcement A_s over a depth of $0.2 h_a$. Reinforcement bars should be anchored to develop 80% of the maximum ultimate force beyond the face of the support.

6.6.1.2 *Shear Capacity*
(a) *Bottom-loaded beams*

Step 1: Check that the shear force V does not exceed the concrete capacity:

$$V < 0.75 b h_a v_u \quad (6.6)$$

where v_u is the shear stress value taken from Table 6 of CP110 (1972) for normal weight concrete or Table 26 for lightweight concrete.

Step 2: Provide hanger bars in both faces to support the bottom loads, using a design stress of $0.87 f_y$. The hanger bars should be anchored to meet the requirements of Clause 2.4.2 of the Guide.

Step 3: Provide nominal horizontal web reinforcement over the lower half of h_a and over the length of the span $0.4 h_a$ measured from each support. This web reinforcement should have an area at least 80% that of the uniformly distributed hanger steel per unit length; it should also satisfy the requirements for bar spacing and the minimum reinforcement percentage in a tension zone, as specified in Section 6.6.1.4 below.

(b) *Top-loaded beams*

The Guide adopts the proven concept of the clear shear span x, as used in eqn (6.1). For uniformly distributed loading, the effective clear shear span x_e is taken to be $l/4$ as proposed earlier by Kong *et al.* (1972).

Step 1: Check that the shear force V does not exceed the lesser of the concrete capacity values given below:

$$V < 2 \frac{b h_a^2}{x_e} v_c \quad \left(\text{for} \frac{h_a}{b} < 4\right) \quad (6.7a)$$

$$V < 1.2 \frac{b h_a^2}{x_e} v_c \quad \left(\text{for} \frac{h_a}{b} \geqslant 4\right) \quad (6.7b)$$

$$V < b h_a v_u \quad (6.8)$$

In eqns (6.7) and (6.8), the effective clear shear span x_e is taken as $l/4$ for uniformly distributed loading; v_c is the shear stress value taken from Table 5 of CP110 (1972) for normal weight concrete or Table 25 for lightweight concrete; v_u is as defined for eqn (6.6).

Step 2: Provide nominal web reinforcement, consisting of horizontal and vertical bars in each face. The amount of reinforcement should not be less than that required for a wall by CP110 (1972): Clauses 3.11 and 5.5; this in effect means at least 0·25% deformed bars in each direction. The horizontal bars should be anchored as links around vertical bars at the edges of the beam; the vertical bars should be anchored around the main bars at the bottom.

6.6.1.3 *Bearing Capacity*
The bearing stresses at supports should not exceed $0·4 f_{cu}$. To estimate the bearing stress at a support, the reaction may be considered uniformly distributed over an area equal to

$$\text{the beam width } b \times \text{the effective support length}$$

the latter to be taken as the actual support length C or $0·2 l_0$ whichever is the lesser.

6.6.1.4 *Crack Control*
The minimum percentage of reinforcement should comply with the requirements of Clauses 3.11 and 5.5 of CP110 (1972). Bar spacings should not exceed 250 mm. In a tension zone, the steel ratio ρ calculated as the ratio of the total steel area to the local area of the concrete in which it is embedded should satisfy the condition

$$\rho > 0·52 \sqrt{f_{cu}}/0·87 f_y \tag{6.9}$$

To control maximum crack widths to within 0·3 mm and 0·1 mm, bar spacings should not exceed those given in Tables 2 and 3, respectively, of the CIRIA Guide.

6.6.1.5 *Web Openings*
Web openings in deep beams will be discussed in Section 6.7 below.

6.6.2 CIRIA's 'Supplementary Rules'
CIRIA's 'supplementary rules' are to be read in conjunction with the 'simple rules'; they cover the design of deep beams generally, including deep beams under concentrated loading, indirect loading and deep beams with indirect supports. Because of space limitation, only single-span beams under top loadings will be dealt with here.

6.6.2.1 *Strength in Bending*
The 'simple rules' described in Section 6.6.1.1 are applicable without modification.

6.6.2.2 Shear Capacity

The ultimate shear capacity V of a deep beam loaded at the top is given by eqn (6.10) below, which applies over the range 0·23–0·7 for x_e/h_a:

$$\frac{V}{bh_a} < \lambda_1 \left[1 - 0\cdot 35 \frac{x_e}{h_a}\right] \sqrt{f_{cu}} + \lambda_2 \sum \frac{100 A_r y_r \sin^2 \theta_r}{bh_a^2} \qquad (6.10)$$

where $\lambda_1 = 0\cdot 44$ for normal weight concrete and 0·32 for lightweight concrete; $\lambda_2 = 1\cdot 95$ N/mm² for deformed bars and 0·85 N/mm² for plain round bars; b = beam thickness; h_a = active height (see eqn 6.3); $A_r, y_r, \theta_r = A, y$ and θ respectively of eqn (6.1) — see Fig. 6.4; x_e = effective clear shear span — see below.

With reference to Fig. 6.4, the effective clear shear span x_e is taken as the least of: (i) the clear shear span x for a load which contributes more than 50% of the total shear force at the support; (ii) $l/4$ for a load uniformly distributed over the whole span; (iii) the weighted average of clear shear spans where more than one load acts and none contributes more than 50% of the shear force at the support. The weighted average is calculated as $\Sigma(V_r x_r)/\Sigma V_r$ where V_r is an individual shear force and x_r its clear shear span; $\Sigma V_r = V$.

The CIRIA equation has been derived from eqn (6.1) by modifying the coefficients C_1 and C_2 of the latter equation so that V as given by eqn (6.10) is about 65% that of eqn (6.1). An inspection of the right-hand side of eqn (6.10) shows that the first term, which represents the concrete contribution, can be tabulated for various values of f_{cu} and x_e/h. Similarly, for a regular pattern of web reinforcement, the second term, which represents the steel contribution, can be tabulated for various steel ratios and x_e/h ratios. Thus, for a beam with an orthogonal web reinforcement, eqn (6.10) can be expressed in a more convenient form for use in design:

$$\frac{V}{bh_a} = \lambda_1 v_x + \beta (v_{ms} + v_{wh} + v_{wv}) \qquad (6.11)$$

where $\lambda_1 = \lambda_1$ in eqn (6.10); $\beta = 1\cdot 0$ for deformed bars and 0·4 for plain round bars; v_x = concrete shear stress parameter which depends on f_{cu} and x_e/h (values are tabulated in Table 4 of the CIRIA Guide); v_{ms} = main steel shear stress parameter which depends on the x_e/h ratio and the main steel ratio (values are tabulated in Table 6 of the CIRIA Guide); v_{wh} (v_{wv}) = horizontal (or vertical) web steel shear parameter which depends on the horizontal (vertical) web steel ratio and the x_e/h ratio (values are tabulated in Table 7 (8) of the CIRIA Guide).

REINFORCED CONCRETE DEEP BEAMS

The design calculations for the shear capacity may now be summarized as follows:

Step 1: If the beam is under concentrated loads, go to Step 2. For uniformly distributed loading, either go to Step 2 or use the more conservative 'simple rules' given earlier.

Step 2: Calculate the shear capacity from eqn (6.10). For a beam with an orthogonal web reinforcement, the same result can be obtained more conveniently from eqn (6.11).
In using eqns (6.10) and (6.11), ignore any web reinforcement which is not within the active height h_a.

Step 3: From the calculations in Step 2, check the total contribution of the (main and web) reinforcement to the shear strength of the beam. If this is less than $0.2V$, increase the web reinforcement to bring the total steel contribution up to at least $0.2V$.

Step 4: Check that the shear capacity of the concrete section is not exceeded:

$$\frac{V}{bh_a} < 1.3\lambda_1 \sqrt{f_{cu}} \qquad (6.12)$$

where λ_1 is defined for eqn (6.10).

6.6.2.3 Bearing Capacity
The limiting bearing stress, set at $0.4f_{cu}$ in the 'simple rules' (see Section 6.6.1.3) may be increased to $0.6f_{cu}$ at end supports and to $0.8f_{cu}$ under concentrated loads, by suitably confining the concrete in the stressed zone. Detailing requirements for the confining reinforcement are given in Clause 3.4.3 of the CIRIA Guide.

6.6.3 Design Example — CIRIA Guide (1977)
Design the reinforcement and the beam thickness for the normal weight concrete deep beam shown in Fig. 6.6, given $f_{cu} = 40$ N/mm², $f_y = 410$ N/mm².

Geometry
From Fig. 6.6: span/depth ratio $\frac{l}{h} = \frac{5250}{4200} = 1.25 < 2$
Hence CIRIA Guide is applicable.
From Fig. 6.6: effective span $l = 5250$ mm
From eqn (6.3): active height $h_a = 4200$ mm

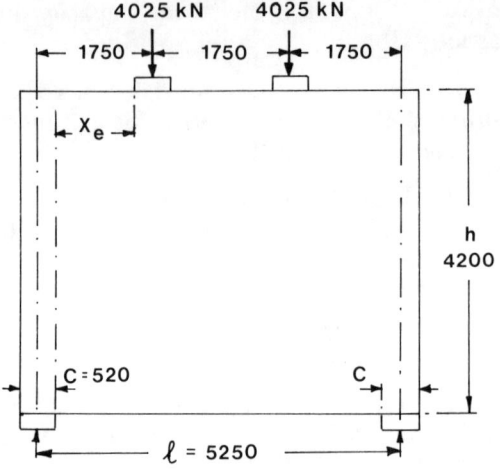

FIG. 6.6. Deep beam in design example. (After Chemrouk, 1984.)

Loading
From CP110 (1972) Clause 2.3.3.1: partial safety factor γ_f for loads = 1·4
Design moment $M = \gamma_f \times 4025 \times 1\cdot75 = 9861$ kNm
Design shear force $V = \gamma_f \times 4025 = 5635$ kN
The beam supports concentrated loads; hence the 'supplementary rules' in Section 6.6.2 are applicable.

Strength in Bending (see Section 6.6.2.1)
Step 1: $l/h_a < 1\cdot5$. Hence eqn (6.4) need not be checked.

Step 2: From eqn (6.5), $z = 0\cdot2l + 0\cdot4h_a = 2730$ mm

$$A_s > \frac{9861 \times 10^6}{0\cdot87 \times 410 \times 2730} = 10\,130 \text{ mm}^2$$

Provide 21 No. 25 mm deformed bars (10 311 mm²).

Step 3: The reinforcement is detailed in 7 layers of 3 bars each, uniformly over a depth of $0\cdot2\,h_a = 840$ mm.

Shear Capacity (see Section 6.6.2.2)
Step 1: Not applicable.

Step 2: Adopt an orthogonal pattern of web reinforcement. From eqn (6.11)

$$\frac{V}{bh_a} < 0.44 v_x + (v_{ms} + v_{wh} + v_{wv})$$

$$f_{cu} = 40 \text{ N/mm}^2; \quad \frac{x_e}{h_a} = \frac{1750 - 520}{4200} = 0.29$$

From CIRIA Guide Table 4: $v_x = 5.68$ N/mm^2
Tentatively assume beam thickness $b = 400$ mm
Main steel ratio $\rho = \dfrac{10\,311}{400 \times 4200} = 0.61\%$
From CIRIA Guide Table 6: $v_{ms} = 1.07$ N/mm^2
Try nominal web reinforcement, 0·25% each way horizontally and vertically.
From CIRIA Guide Tables 7 and 8: $v_{wh} = 0.22$ N/mm^2 and $v_{wv} = 0$.
Hence total shear capacity is

$$(0.44 \times 5.68 + 1.07 + 0.22 + 0) \times 400 \times 4200 \times 10^{-3} = 6366 \text{ kN}$$

$V = 5635$ kN < 6366 kN O.K.

Step 3:

$$(v_{ms} + v_{wh} + v_{wv}) bh_a = (1.07 + 0.22 + 0) \times 400 \times 4200 \times 10^{-3}$$
$$= 2167 \text{ kN}$$
$$0.2 V = 0.2 \times 5635 = 1127 \text{ kN} \quad \text{O.K.}$$

Step 4: From eqn (6.12)

$$\frac{5635 \times 10^3}{400 \times 4200} < 1.3 \times 0.44 \sqrt{40}$$

$$3.35 < 3.62 \quad \text{O.K.}$$

Bearing Capacity (see Section 6.6.2.3)
From Fig. 6.6:
Actual support length $C = 520$ mm.

$$0.2 l_0 = 0.2 \times (5250 - 520) = 946 \text{ mm} < 520 \text{ mm}$$

Hence effective support length = 520 mm

$$\frac{V}{bc} < 0.6 f_{cu}$$

$$\frac{5630 \times 10^3}{b \times 520} < 0.6 \times 40$$

$$b > 451 \text{ mm}$$

Revise b to 460 mm.

The detailing is shown in Fig. 6.7. For further discussions and design examples, covering both bottom loading and a combination of top and bottom loading, see Chemrouk (1984).

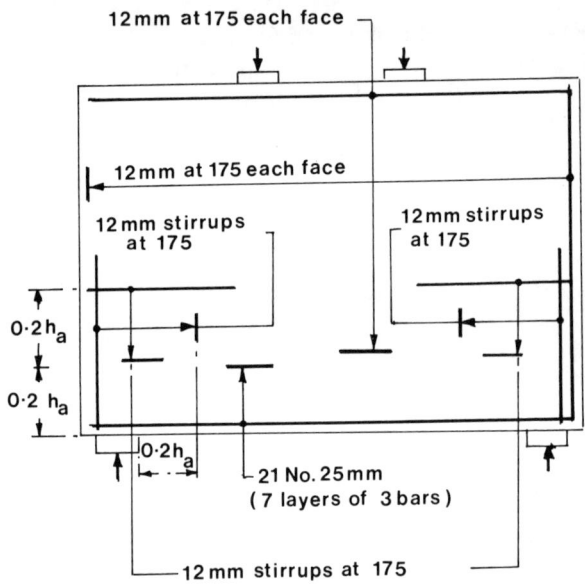

FIG. 6.7. Reinforcement details in design example. (After Chemrouk, 1984.)

6.7 WEB OPENINGS IN DEEP BEAMS

Deep beams with web openings are not yet covered by the current codes of practice such as ACI 318 (1983), the CEB–FIP Model Code (1978) and BS 8110 (1985). Even the comprehensive CIRIA Guide (1977) is cautious and restrictive in its recommendations; this is because of the scarcity of information available at the time of its preparation (Kong and Sharp, 1973).

The exact analysis of deep beams with web openings presents formidable problems. However, from the observed behaviour of such beams (Kong and Sharp, 1973; 1977; Kong and Kubik, 1986; Kong et al., 1978) it is possible to formulate a structural idealization which focuses attention on the main features of the complex load-transfer mechanism, ignoring lesser quantities.

The structural idealization is shown in Fig. 6.8. The applied load is transmitted to the support mainly by a lower path ABC and partly by an upper path AEC. Thus, the larger the angle ϕ, the more effective is the lower path; similarly, the upper path becomes more effective as ϕ' increases. In the absence of the opening, the upper and lower paths become one, which is the natural load path joining the loading and reaction points; in this case the ultimate shear strength is of course given by eqn (6.1). The structural idealization shows that if the opening is small or so located as not to interfere significantly with the natural load path, eqn (6.1) will give a reasonable estimate of the shear strength. If the opening interrupts the natural load path, eqn (6.1) is modified as

$$V = C_1 \left[1 - 0.35 \frac{k_1 x}{k_2 h} \right] f_t b k_2 h + \sum \lambda C_2 A \frac{y_1}{h} \sin^2 \theta_1 \qquad (6.13)$$

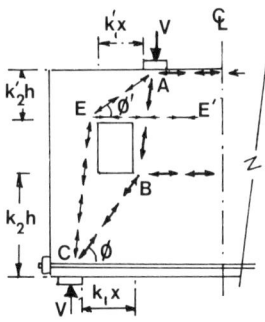

FIG. 6.8. Structural idealization for deep beams with web openings.

where k_1, k_2 = coefficients defining the position of an opening (Fig. 6.8); y_1 = depth at which a typical bar intersects a potential critical diagonal crack in a deep beam with openings, idealized as the line EA or CB in Fig. 6.9; λ = a coefficient equal to 1·5 for web bars and 1·0 for main bars. The other symbols are as defined in eqn (6.1).

Equation (6.13) has been checked against ultimate load tests on deep beams incorporating an unusually wide range of web openings and reinforcement types (Kong and Sharp, 1977; Kong *et al.*, 1978; Kong and Kubik, 1986). The more recent tests include eight large deep beams measuring 2 m × 1·8 m × 0·25 m and weighing about 4·5 tonnes each (Kong and Kubik, 1986); these are believed to be among the largest of their kind ever tested to destruction in a laboratory.

Intending users of eqn (6.13) should study the comments on eqn (6.1) given earlier in this chapter; they should also study the design hints given by Kong and Sharp (1977). For design examples, see Reynolds and Steedman (1981) and Chemrouk (1984).

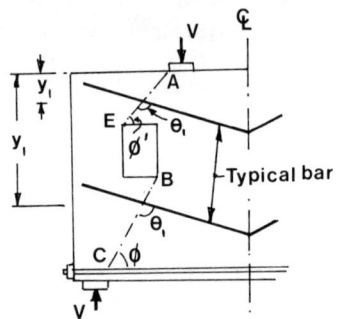

FIG. 6.9. Meanings of symbols in eqn (6.13).

6.8 CONCLUDING REMARKS

Practical research on the ultimate load behaviour of reinforced concrete deep beams can be said to begin with the experiments of de Paiva and Siess (1965) and Leonhardt and Walther (1966). Since then, a substantial amount of work has been done; apart from the works referred to above, mention should be made of those by Kubik (1980), Singh et al. (1980), Regan and Hamadi (1981), Roberts and Ho (1982), Smith and Vantsiotis (1982) and Subedi (1983). However, the effects of repeated loading (Kong and Singh, 1974) and bottom loading are still largely unknown. Many engineers still have doubts on the applicability of plastic methods (Kong and Charlton, 1983; Kong et al., 1983a) to deep-beam analysis and design; this has not been helped by the confusion over plastic theory itself, and the difficulties are unlikely to be resolved until the currently widespread misunderstanding of the Principle of Virtual Work can somehow be cured (Kong et al., 1983b). The instability and buckling of slender deep beams remains a difficult topic and information is hard to find in the literature (Tang, 1986; Wong, 1986). The CIRIA Guide (1977) gives some helpful advice on design procedures but warns that 'There is no experimental evidence to substantiate these procedures'. The Author and his colleagues have been studying this topic in recent years and will be providing some experimental evidence soon (Kong et al., 1986).

ACKNOWLEDGEMENTS

The Author wishes to thank his research students, Mr C. W. J. Tang and Mr H. H. A. Wong, for valuable discussions and careful proof reading, and to thank his former research student, Mr M. Chemrouk, for providing the design example in Section 6.6.3.

REFERENCES

ACI 318-71 (1971) *Building Code Requirements for Reinforced Concrete*, American Concrete Institute, Detroit, 78 pp.
ACI 318-83 (1983) *Building Code Requirements for Reinforced Concrete*, American Concrete Institute, Detroit, 111 pp.
ALBRITTON, G. E. (1965) *Review of Literature Pertaining to the Analysis of Deep Beams*. Technical Report No. 1-701, US Army Engineer Waterways Experiment Station, Vicksburg, Miss., 80 pp.
BESSER, I. I. and CUSENS, A. R. (1984) Reinforced concrete deep beam panels with high depth/span ratios. *Proc. ICE, Part 2*, 77, 265-78.
BS 8110 (1985) *The Structural Use of Concrete*, British Standards Institution, London, Part 1, 121 pp.
C & CA (1969) Bibliography on Deep Beams, Library Bibliography No. Ch. 71(3/69), Cement & Concrete Association, London, 8 pp.
CEB-FIP (1970) *International Recommendations for the Design and Construction of Concrete Structures. Appendix 3: Deep Beams*, English Edition, Cement & Concrete Association, London.
CEB-FIP (1978) *Model Code for Concrete Structures*, English Edition, Cement & Concrete Association, London, p. 200.
CHEMROUK, M. (1984) Design of Reinforced Concrete Deep Beams. MSc Thesis, University of Newcastle upon Tyne, 202 pp.
CIRIA GUIDE 2 (1977) *The Design of Deep Beams in Reinforced Concrete*, Ove Arup & Partners and Constructional Industry Research & Information Association, London, 131 pp.
COATES, R. C., COUTIE, M. G. and KONG, F. K. (1980) *Structural Analysis*, Van Nostrand Reinhold (UK) Ltd, 2nd Edition, 579 pp.
COPE, R. J., SAWKO, F. and TICKELL, R. G. (1982) *Computer Methods for Civil Engineers*, McGraw Hill (UK) Ltd, 361 pp.
CP110 (1972) *The Structural Use of Concrete*, British Standards Institution, London, Part 1, 154 pp.
DE PAIVA, H. A. R. and SIESS, C. P. (1965) Strength and behaviour of deep beams in shear. *Proc. ASCE*, 91 (ST5), 19-41.
KONG, F. K. and CHARLTON, T. M. (1983) The fundamental theorems of the plastic theory of structures. In *Proc. M.R. Horne Conf. on Instability and Plastic Collapse of Steel Structures, Manchester* (Editor J. L. Morris), Granada Publishing, London, pp. 9-15.
KONG, F. K. and EVANS, R. H. (1980) *Reinforced and Prestressed Concrete*, Van Nostrand Reinhold (UK) Ltd, 2nd Edition, pp. 163-5 and 178-9.

KONG, F. K. and KUBIK, L. A. (1986) Large scale tests on reinforced concrete deep beams with web openings. Paper in preparation.
KONG, F. K. and SHARP, G. R. (1973) Shear strength of lightweight reinforced concrete deep beams with web openings. *Structural Engineer*, **51**, 267-75.
KONG, F. K. and SHARP, G. R. (1977) Structural idealization for deep beams with web openings. *Magazine of Concrete Research*, **29**, 81-91.
KONG, F. K. and SINGH, A. (1972) Diagonal cracking and ultimate loads of lightweight concrete deep beams. *Proc. ACI*, **69**, 513-21.
KONG, F. K. and SINGH, A. (1974) Shear strength of lightweight concrete deep beams subjected to repeated loads. In *Shear in Reinforced Concrete*, SP42, American Concrete Institute, Detroit, pp. 461-76.
KONG, F. K., ROBINS, P. J., SINGH, A. and SHARP, G. R. (1972) Shear analysis and design of reinforced concrete deep beams. *Structural Engineer*, **50**, 405-9.
KONG, F. K., ROBINS, P. J. and SHARP, G. R. (1975) The design of reinforced concrete deep beams in current practice. *Structural Engineer*, **53**, 173-80.
KONG, F. K., SHARP, G. R., APPLETON, S. C., BEAUMONT, C. J. and KUBIK, L. A. (1978) Structural idealization of deep beams with web openings: further evidence. *Magazine of Concrete Research*, **30**, 89-95.
KONG, F. K., EVANS, R. H., COHEN, E. and ROLL, F. (1983a) *Handbook of Structural Concrete*, Pitman, London, 1968 pp.
KONG, F. K., PRENTIS, J. M. and CHARLTON, T. M. (1983b) Principle of virtual work for a general deformable body — a simple proof. *Structural Engineer*, **61A**, 173-9.
KONG, F. K., GARCIA, R. C., PAINE, J. M., WONG, H. H. A., TANG, C. W. J. and CHEMROUK, M. (1986) Strength and stability of slender concrete deep beams. *Structural Engineer*, in press.
KUBIK, L. A. (1980) Predicting the strength of reinforced concrete deep beams with web openings. *Proc. ICE, Part 2*, **69**, 939-58.
LEONHARDT, F. and WALTHER, R. (1966) *Deep Beams*, Bulletin 178, Deutscher Ausschuss für Stahlbeton, Berlin, 159 pp. (English Translation: CIRIA, London, 1970.)
REGAN, P. E. and HAMADI, Y. D. (1981) *Concrete in the Oceans. Part 1: Reinforced Concrete Deep Beams with Thin Webs*, Cement and Concrete Association, London, pp 1-37.
REYNOLDS, C. E. and STEEDMAN, J. C. (1981) *Reinforced Concrete Designer's Handbook*, Viewpoint Publications, London, 9th Edition, pp. 402-3.
ROBERTS, T. M. and HO, N. L. (1982) Shear failure of deep fibre reinforced concrete beams. *International Journal of Cement Composites and Lightweight Concrete*, **4**, 145-52.
ROBINS, P. J. and KONG, F. K. (1973) Modified finite element method applied to reinforced concrete deep beams. *Civil Engineering & Public Works Review*, **68**, 963-6.
SINGH, R., RAY, S. P. and REDDY, C. S. (1980) Some tests on reinforced concrete deep beams with and without opening in the web. *Indian Concrete Journal*, **54**, 189-94.
SMITH, K. N. and VANTSIOTIS, A. S. (1982) Shear strength of deep beams. *Proc. ACI*, **79**, 201-13.

SUBEDI, N. K. (1983) The behaviour of reinforced concrete flanged beams with stiffeners. *Magazine of Concrete Research*, **35**, 40–54.

TANG, C. W. J. (1986) PhD Thesis in preparation, University of Newcastle upon Tyne.

WONG, H. H. A. (1986) PhD Thesis in preparation, University of Newcastle upon Tyne.

ZIENKIEWICZ, O. C. (1977) *The Finite Element Method,* McGraw Hill (UK) Ltd, 3rd Edition, 787 pp.

Chapter 7

COLUMN-SUPPORTED SHEAR WALLS

LADISLAV CERNY and ROBERTO LEON
Department of Civil and Mineral Engineering, University of Minnesota, Minneapolis, Minnesota, USA

SUMMARY

It is common, for architectural reasons, to desire unobstructed spaces in the lower floors of multistorey buildings. If the structural system utilized to carry either the gravity load or lateral loads consists of shear walls, it becomes necessary to support these walls on widely spaced columns. The discontinuity in geometry at the lowest level results in a large change in the strength and stiffness of the structure near its critical section, and presents the designer with several problems. The main problems relate to the evaluation of the loads acting on and the design of the transfer, or lintel, beam below the wall, the detailing of the wall-to-beam and beam-to-column connections, as well as the design of the columns themselves. This chapter discusses the methodologies available today to analyze this type of structure, compares their accuracy, and proposes some practical solutions to the problems encountered.

NOTATION

A_i Area of each individual wall
b Beam support width (support column width)
C Load concentration factor over the support (solid wall)
c'' Load concentration factor at center of transfer beam
d Half-depth of transfer beam
f Extent of vertical force on transfer beam
H Wall height from the centroid of transfer beam
h Storey height

I_b Moment of inertia of transfer beam
I_c Moment of inertia of wall section
L Wall span
l Center-to-center distance between walls
M Moment in the support beam
s Clear width of opening between walls
T Axial force in the transfer beam
T_s Axial force from secondary arch
t Wall thickness
V Shear in the support beam
W Total vertical load at the transfer beam level

7.1 INTRODUCTION

Since the lower storeys of large buildings are often used for public facilities such as restaurants, stores, or theatres, it is common to require large, unobstructed spaces in these areas. If the structural system devised for the rest of the building contains large shear walls, it becomes necessary to support them on widely spaced columns at the lower storeys (Fig. 7.1). The structural engineer is then faced with the task of

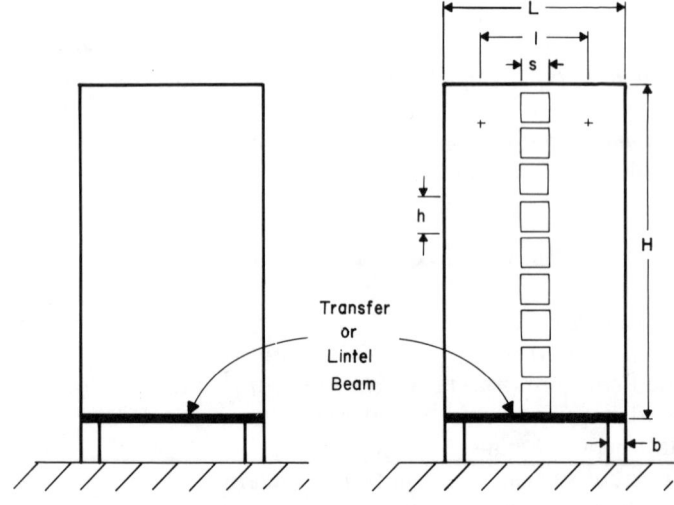

FIG. 7.1. Shear walls supported on columns.

determining the design loads, and detailing a suitable section to transfer such loads from the walls to the bottom storey columns. The problem is complicated by the fact that typically there will be openings in the walls to accommodate corridors, and thus the analysis must account for their effect on the load distribution.

7.1.1 Description of the Problem
Shear walls supported on columns are typically used when the lateral loads are relatively small, and where no appreciable dynamic effects need to be considered in the design. Given the drastic changes in stiffness and strength that this system exhibits near the critical section, its use has been strongly discouraged in seismic areas (Paulay, 1975) when the wall is a part of the structural system carrying lateral loads. This has limited its use mostly to non-seismic areas, and even there it is not common. It is generally considered on a case-by-case basis and no general design guidelines exist for such structures.

7.1.1.1 *Gravity Loading*
When a shear wall supported on columns carries a gravity load only, a pronounced arch effect takes place near the bottom of the wall in order to transfer the stresses from the middle of the wall to the columns located at the extremes (Fig. 7.2). This arch section results in a very large

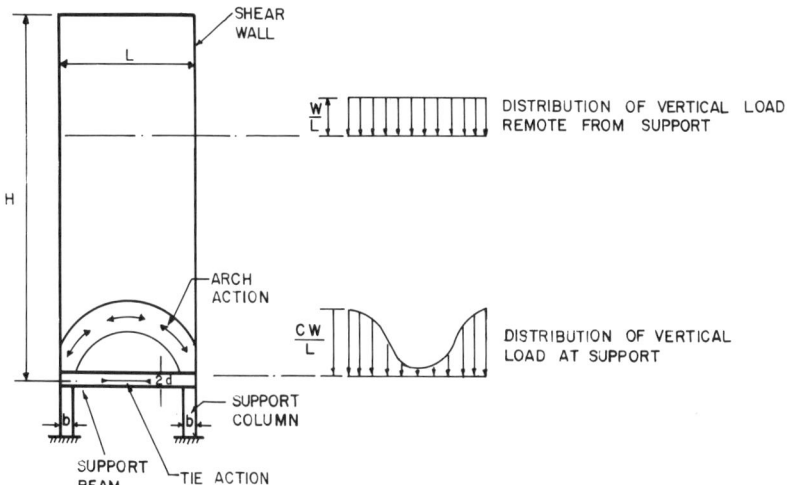

FIG. 7.2. Arch action in the shear wall supported on columns.

axial tensile force, in addition to shear and bending moments, in the transfer beam which acts as the bottom tie in a tie arch structure. Utilizing the tied arch analogy for the analysis will result in either a very deep or a heavily reinforced transfer beam, or a combination thereof. In addition, such analysis assumes that the axial force must be transferred as a horizontal shear between the wall and the transfer beam, and thus the boundary between the wall and transfer beam must be carefully detailed. The horizontal forces and moments from the transfer beam must then be combined with those from the arch action to obtain the design forces for the columns. The columns required to carry these loads are very wide and heavily reinforced, presenting problems both from the design and construction standpoints.

7.1.1.2 *Lateral Loads*

As previously noted, the use of shear walls supported on columns is not encouraged where lateral loads are high or large dynamic effects are present because of the potential for overturning. Experiences from past earthquakes in developing countries indicate very poor performance, particularly for the common case of masonry walls resting on columns. In many cases failures can be attributed to poor design, construction errors, or poor materials, but dynamic analysis of this type of structure indicates that it would be difficult to eliminate column compression and shear failures under significant ground motions. The failure of the lower storey columns in the Macuto Sheraton during the 1967 Caracas earthquake (Esteva, 1980), and the recent failure of the Imperial County Services Building (Sheperd and Plunkett, 1983) are testimony to the poor performance of this structural system under seismic loading.

In areas where the lateral loads are not high, i.e. governed by wind loads, the system can be satisfactorily used if the analysis and design are carefully carried out. This is particularly true for systems in which only a limited number of the walls are supported on columns, while the majority of the walls are continuous to the foundation. For this case it is likely that the distribution of lateral loads will not be uniform, and therefore the walls on columns will not be required to carry large lateral loads. The application of lateral load to a shear wall on column supports results in large overturning forces in the wall, as well as large axial and shear forces in the supporting columns. It should be noted that lateral loads do not tend to govern the design of the transfer beams, since this loading does not introduce significant axial loads in the beams. The design of the columns, on the other hand, is governed by a combination of gravity and lateral loads, and therefore it becomes important to utilize an analytical approach that can handle both kinds of loading.

7.2 ANALYSIS OF SHEAR WALLS SUPPORTED ON COLUMNS

The analysis of shear walls on support columns was originally developed as a variation of the case of shear walls with openings. Some of the earlier work was carried out by MacLeod (1967), Schwaighofer and co-workers (1967, 1969), Smith (1970), and others. Good summaries of the early development of the techniques are given by Coull and Smith (1967) and MacLeod (1971). Over the years four different approaches to the problem have been proposed. For discussion purposes, the four approaches will be labelled as the approximate method, the frame idealization approach, the finite element approach, and the experimental approach.

7.2.1 Approximate Approach
A preliminary design approach to the problem of determining the forces on the transfer or lintel beam of a shear wall supported by columns was first proposed by Green et al. (1973). This approach was proposed for shear walls without openings, with height to width ratios greater than 1·5, and was based on a finite element parameter study. The magnitude of the forces in the transfer beam were derived on the basis of a simplified stress distribution at the bottom of the wall, and simple equilibrium equations (see Fig. 7.3). The method was modified empirically to account for openings in the wall, and resulted in forces in the beams that depended on the depth and stiffness of the transfer beam, the effective width of the supporting columns, and the size and positioning of the opening. The first step in this procedure is to compute a load concentration factor C, representing the ratio of actual stress above the support column versus the average vertical stress for the wall, based on the in-plane stiffness of the wall relative to the beam. A plot of this concentration factor is shown in Fig. 7.4. From this distribution and some simplifying assumptions, the following forces are derived:

$$\frac{T}{W} = \frac{1}{4} - \frac{b}{L}\left(1 - \frac{Cb}{L}\right) \qquad (7.1)$$

$$V_{\text{support}} = \frac{W}{2}\left(1 - \frac{2Cb}{L}\right) \quad (7.2)$$

$$M_{\text{support}} = \frac{WL}{12}\left[C\left(\frac{f}{L}\right)^2\left(2 - \frac{f/L}{(1-2b/L)}\right) + \frac{5}{8}c''\left(1 - 2\frac{b}{L}\right)^2\right] - \frac{2}{3}Td \quad (7.3)$$

$$M_{\text{center}} = -\frac{WL}{12}\left[\frac{C(f/L)^3}{(1-2b/L)} + \frac{3}{8}c''\left(1 - 2\frac{b}{L}\right)^2\right] + \frac{Td}{3} \quad (7.4)$$

If $\left(\frac{1}{C} - 2\frac{b}{L}\right) \leq \left(\frac{1}{2} - \frac{b}{L}\right)$ then $c'' = 0$ and $\frac{f}{L} = \left(\frac{1}{C} - 2\frac{b}{L}\right)$.

Otherwise $c'' = \dfrac{2 - C(1 + 2b/L)}{(1 - 2b/L)}$ and $\dfrac{f}{L} = \left(\dfrac{1}{2} - \dfrac{b}{L}\right)$.

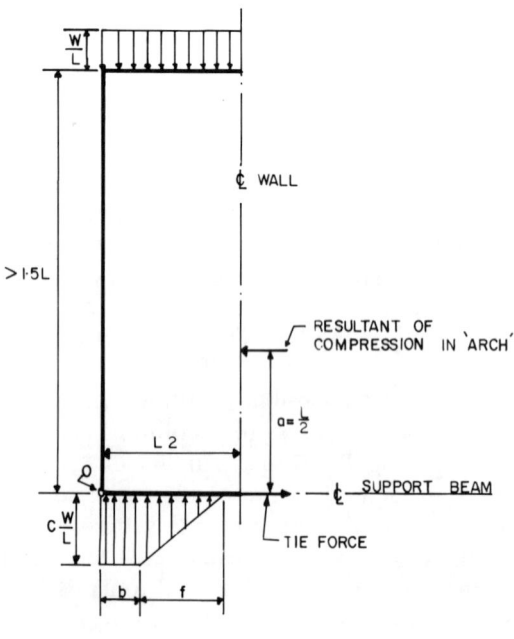

FIG. 7.3. Approximate method of shear wall analysis. (a) Idealized force action.

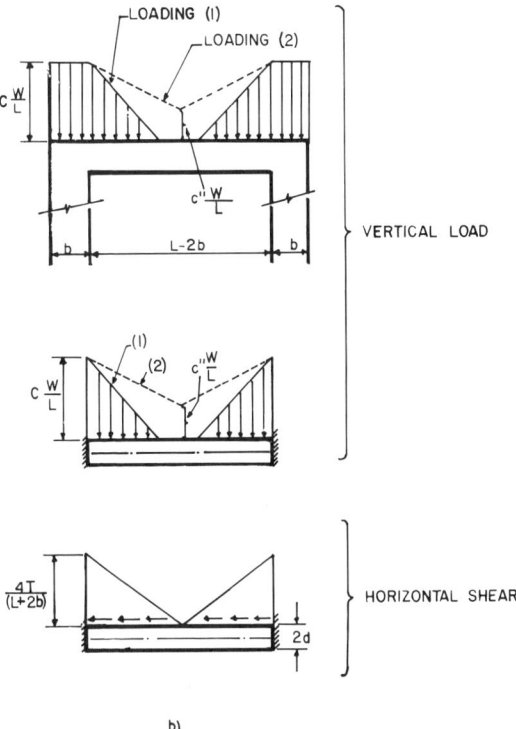

FIG. 7.3—contd. (b) Force action for support beam.

To account for openings in the walls:

$$C_{\text{(openings)}} = C_{\text{(without openings)}} (1 - \{s/L\}) \qquad (7.5)$$

where s is the width of the openings.

Comparison of the values obtained from this procedure with those from an instrumented building resulted in an overestimation by a factor of 2 to 3 of the axial force in the transfer beam if the tensile capacity of the concrete was ignored. If the tensile capacity of the concrete was accounted for, the procedure underestimated the force in the transfer by about a factor of 2 (Fig. 7.5). While the use of the equations without the tensile capacity proved conservative and therefore adequate for design purposes, the procedure cannot be considered to be accurate analytically.

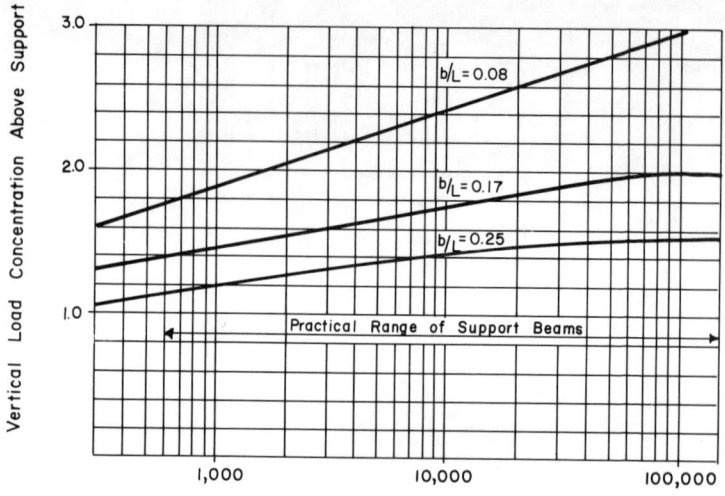

FIG. 7.4. Load concentration factor.

The procedure discussed above is limited to gravity loads only. For the case of lateral loads, MacLeod (1971) proposed the idealization shown in Fig. 7.6 to account for the fact that the proportion of the lateral load carried by the wall may be affected by the stiffness discontinuity at the first floor. In this procedure the floor slabs are assumed to be rigid, and compatibility of deformations is used to apportion the lateral load to the different frames in the structure. This procedure can provide the forces in the columns but not the axial force in the transfer beam.

7.2.2 The Frame Idealization

To take advantage of the availability of computer programs based on the direct stiffness method and to provide more accurate solutions to the problem of shear walls with openings, MacLeod and Green (1973) introduced the frame idealization technique. In this case the walls are modelled as a frame consisting of very rigid or fully rigid horizontal elements representing the **walls**, linked by connecting beams and columns having a finite stiffness (Fig. 7.7). The factors affecting the resultant forces are the axial and bending stiffness of the support beam, the bending stiffness of the columns and beams, the size and

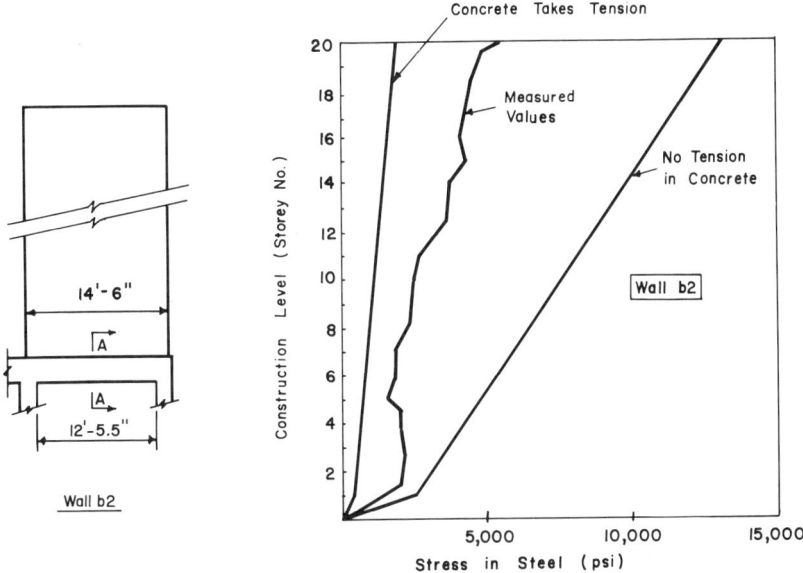

FIG. 7.5. Comparison of the results from the approximate method with the field measurements (wall b2).

positioning of the openings, the width of the supporting columns, and the type of loading. Tables 7.1 and 7.2 show some of the comparisons for two buildings studied. The study was based on varying a parameter α, defined as

$$\alpha = \sqrt{\frac{12I_b}{hb^3}\left(\frac{l^2}{\Sigma I_c} + \frac{1}{A_1} + \frac{1}{A_2}\right)} \quad (7.6)$$

Thus $\alpha H = 4$ represents a wall where the connecting beams are flexible, and $\alpha H = 15$ represents a wall where the connecting beams are stiff. Comparison of all results gave good correlation except for the maximum axial force in the transfer beam. In this case the maximum value for this force occurs below the stiffer wall section as opposed to at the opening as the frame idealization would predict. To correct for this discrepancy the formation of a secondary arch is hypothesized, as shown in Fig. 7.8, and a modification of the axial force in the beam is proposed as follows:

T(total) = T(primary arch = axial force from frame analysis) +

T(secondary arch = shear in support beam from frame analysis)

(7.7)

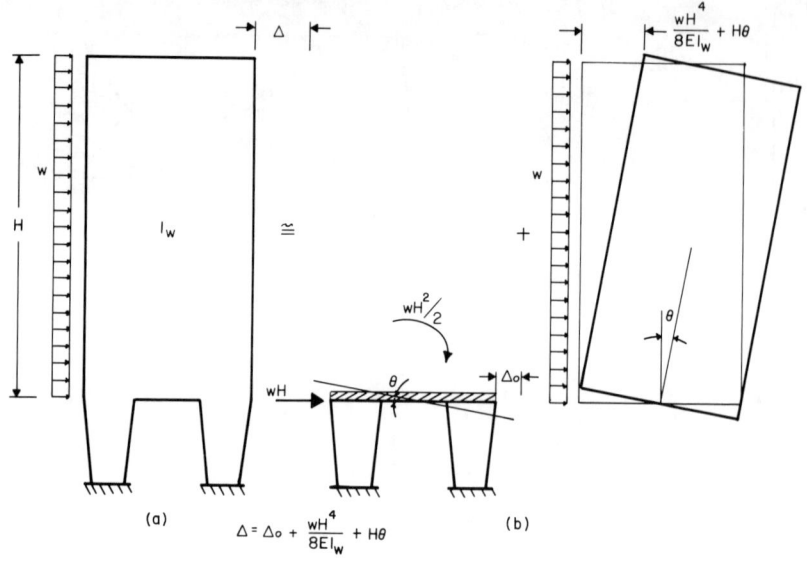

Shear Wall with Column Supports

FIG. 7.6. Approximate evaluation of forces in support columns.

TABLE 7.1
DETAILS OF FRAMES ANALYZED

	$I(m^4)$	$A(m^2)$
Connecting beams		
$\alpha H = 4$	0·000 1	0·087 5
$\alpha H = 15$	0·001 823	0·087 5
Support beam	0·003 125	0·015
Support column	0·006 25	0·3

This more sophisticated method, however, tends to overestimate the axial force in the transfer beam and, while conservative, cannot be regarded as a good predictor of the axial force in the lintel beam. It should be noted that this procedure is susceptible to numerical instabilities given the large differences in stiffnesses for some of the elements; if a computer is used to obtain the solution, it becomes difficult to determine whether any instabilities have occurred.

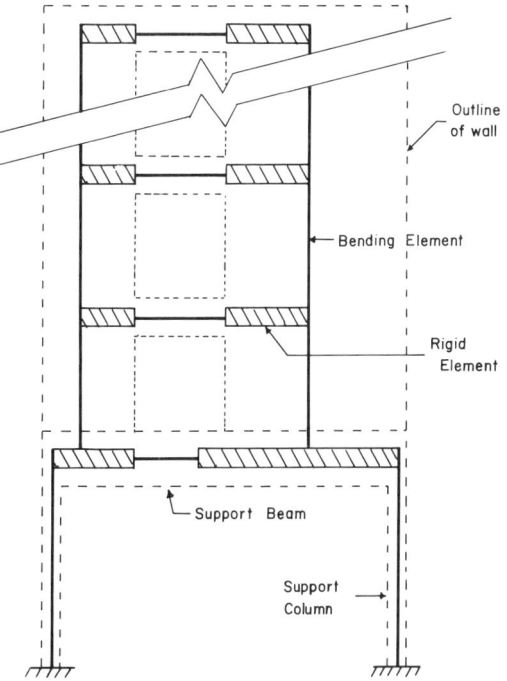

FIG. 7.7. Typical frame idealization.

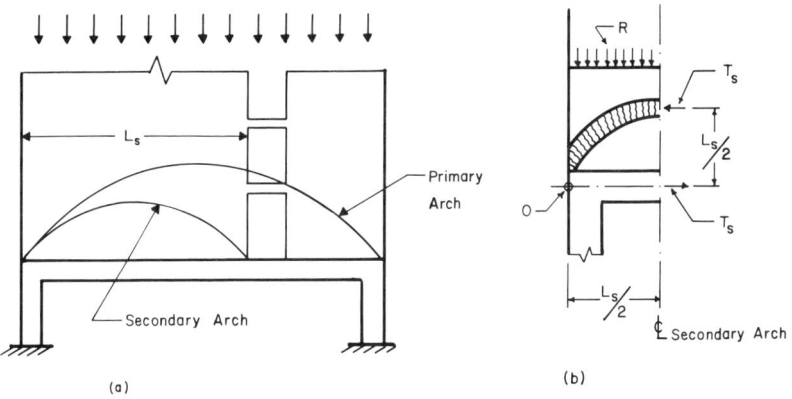

FIG. 7.8. Arch action for correction of axial force in the transfer beam. (a) Arch action; (b) free body diagram of secondary arch.

TABLE 7.2.
COMPARISON OF FORCE ACTIONS BY FRAME IDEALIZATION AND FINITE ELEMENT IDEALIZATION

	$\alpha H = 15$				$\alpha H = 4$			
	Symmetrical		Off-center		Symmetrical		Off-center	
(kN)	Frame	Finite element	Frame	Finite element	Frame	Finite element	Frame	Finite element
Axial force in support beam at opening	+125·3	+104·6	+78·2	+63·9	+127·1	+10·53	+82·1	+69·3
Maximum axial force in support beam	+125·3	+104·6	+178·8[a]	+152·8	+127·1	+105·3	237·4[a]	+203·9
Shear force in support beam at opening	0	0	−100·6	−91·1	0	0	−155·3	−142·1
Shear in connecting +1 beams +2 +3	0 0 0	0 0 0	−75·1 −31·4 −11·9	−64·5 −31·5 −14·3	0 0 0	0 0 0	−19·1 −13·5 −10·1	−19·4 −15·8 −13·6

Gravity load

COLUMN-SUPPORTED SHEAR WALLS

Lateral load	Axial force in support beam at opening		±2.6	±2.2	±20.8	±19.9	±2.6	±3.9	±12.5	±5.7
	Maximum axial force in support beam			See note below[b]						
	Shear force in support beam at opening		±26.1	±19.8	±60.0	±44.3	±79.5	±66.5	±106	±92.8
	Shear force in connecting beams	+1	±30.7	±34	±26.4	±31.9	±12.5	±14	±12	±14.4
		+2	±33.3	±36.5	±27.5	±31.8	±17	±17.6	±14.5	±16.5
		+3	±31.7	±29.5	±26.6	±25.5	±19.5	±18.9	±15.5	±17.0

[a] By approximate formula.
[b] Lateral load on beam-supported shear walls does produce axial force in the support beam which is generally insignificant. In some cases, however, where the support beam is stiff in comparison with the connecting beams this effect may be as much as 20% of the maximum axial force due to gravity load. The frame idealization cannot predict this force action.

7.2.3 The Finite Element Method (FEM)

The advent of mainframe computers in the early 1970s and the subsequent development of the finite element method have made FEM the preferred solution scheme for the problem of shear walls on support columns. Although expensive and time-consuming insofar as data preparation is concerned, the finite element method presents the only sufficiently accurate solution to the determination of the forces in this type of structure. From the design standpoint, the expense of computer solutions is more than justified since it can result in substantial savings in materials and construction costs as well as design time.

The accuracy of the finite element solution depends heavily on the type of element chosen and the refinement of the mesh. Utilizing simple elements (plain stress, for example) requires a fairly fine mesh to achieve satisfactory results, while some of the newer higher-order elements would probably require a far lower number for equal accuracy.

7.2.4 Experimental Approaches

As with many structures difficult to model in everyday design, the problem of shear walls on support columns can be attacked from the experimental standpoint with the use of small-scale models. Since only axial and bending forces are of interest, plastic models and photo-elasticity can be easily used to predict the forces in the lowest transfer beam. The approach has been used (Cerny, 1979) to verify that the smaller forces predicted in the transfer beam by the finite element method are closer to the real values than the results of the frame idealization or the approximate methods previously discussed. Similar investigations can be carried out on models constructed from other materials and instrumented with resistance strain gauges.

7.2.5 Comparison of Design Approaches

To compare the accuracy of the last three methods outlined above, a comparative analysis was carried out by Cerny (1979) on the structure shown in Fig. 7.9. This was a structure designed and built in the early 1970s and which has shown satisfactory performance under service loads. The structure was subjected to two different loading conditions. Loading 1 corresponded to a wind force of $1 \cdot 44 \, kN/m^2$ (30 psf) uniformly distributed along the height of the structure. Loading 2 corresponded to a combination of dead and live loads, again in accordance with the applicable UBC specifications. Loading 2

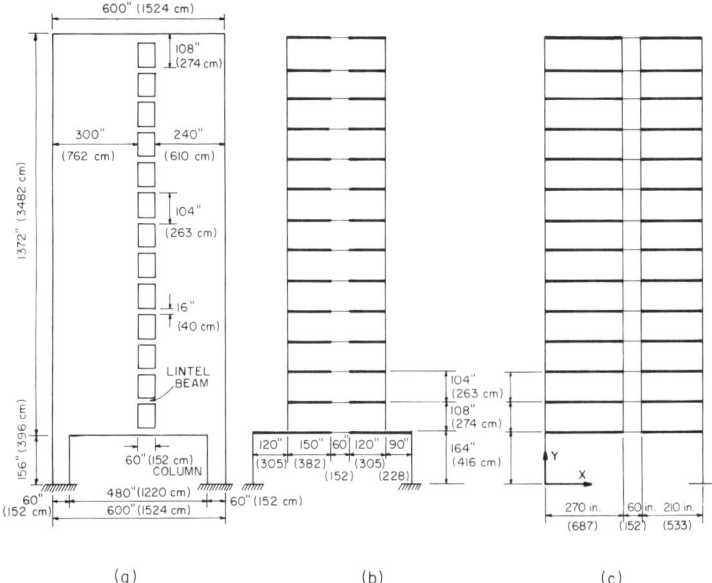

FIG. 7.9. (a) Hybrid coupled shear wall; (b) equivalent frame; (c) equivalent frame.

produced the largest forces in the transfer beam, while a combination of loads 1 and 2 produced the largest moments in the columns. It was clear that the main parameter governing the magnitude of the forces was the relative stiffness of the transfer beam versus the columns, and this was selected as the main variable in the investigation.

Two different frame idealization models and two different meshes for the FEM were used in the study. One of the frame idealizations was based on the MacLeod approach, with the columns located in the middle of the wall width (Fig. 7.9b), while the other was based on keeping the wall elements directly above the support columns (Fig. 7.9c). The FEM was used on models such as those shown in Fig. 7.10, with the number of elements varying from 413 to 2000. The results were verified by a photoelastic study on models constructed from hard and soft urethan and polyester. The measured whole and half fringes due to gravity loads are shown in Fig. 7.11.

The results of these comparisons are shown in Tables 7.3 and 7.4. In Table 7.3 the results of loading case 2 are shown for the axial force in the transfer beam. The FEM results, which correlate very well with the

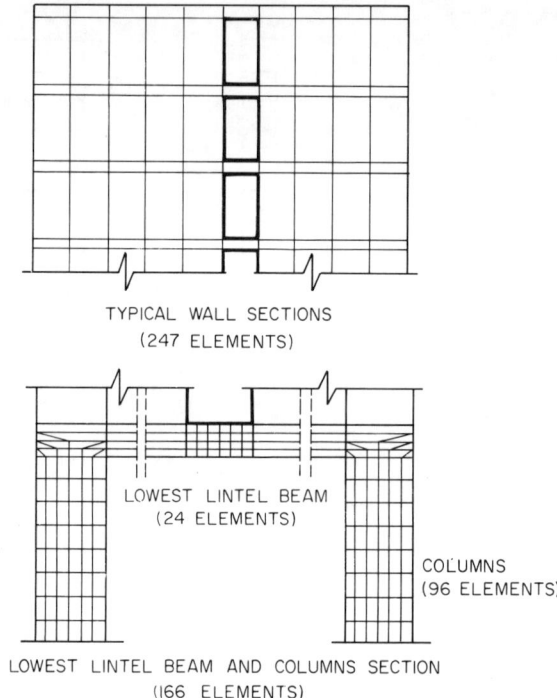

FIG. 7.10. Finite element model.

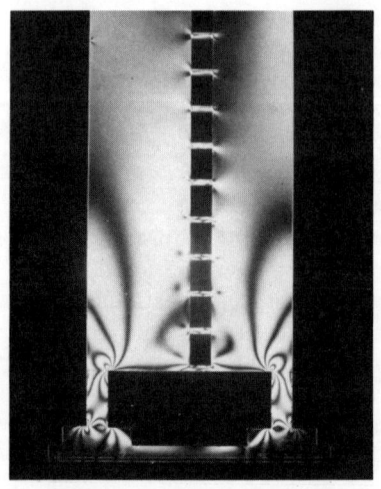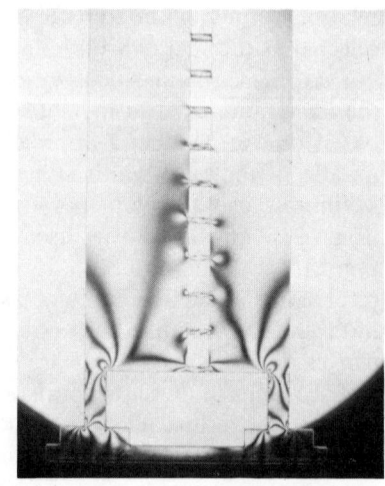

FIG. 7.11. Photoelastic model. (left) Whole fringes; (right) half fringes.

TABLE 7.3
AXIAL FORCE IN THE LOWEST LINTEL BEAM, kips (1 kip = 4·45 kN) LOADING 2

	Case 1	Case 2	Case 3	Case 4	Case 5
Cross-sectional properties of columns and lowest lintel beam: in (cm)	Column 8 × 60 (20 × 152) Beam 8 × 16 (20 × 40)	Column 16 × 60 (40 × 152) Beam 8 × 16 (20 × 40)	Column 16 × 60 (40 × 152) Beam 12 × 16 (30 × 40)	Column 16 × 60 (40 × 152) Beam 16 × 16 (40 × 40)	Column 16 × 60 (40 × 152) Beam 24 × 16 (60 × 40)
Frame analogy	374	305	365	404	430
Frame analogy	—	—	—	478	523
Finite element model 413 elements	205	173	200	220	235
Finite element model 2 000 elements	185	161	182	198	212
Photoelastic model from polyester resin	205	Not tested	195	200	Not tested

photoelastic studies, show axial forces about one-half as large as those from the frame idealization. In Table 7.4 the results for the moments in the columns are shown; here the differences are smaller, in the order of 25%.

TABLE 7.4
MOMENTS IN COLUMNS, LOADING 2 kips-in (1 kip-in = 16·3 kNm)

Column	Case 1	Case 2	Case 3	Case 4	
Left	8 910 5 859	15 615 9 958	13 842 8 609	12 683 7 734	Frame analogy
Right	8 692 6 077	15 078 10 495	13 375 9 076	12 280 8 137	
Left	8 555 5 220	11 585 7 640	11 520 7 125	11 405 6 825	Finite elements
Right	8 455 5 305	11 390 7 800	11 310 7 380	11 195 6 435	
Left	8 880 5 400				
Right	8 200 5 350	Photoelastic model from polyester			

Limited strain measurements were made on the columns and the lowest lintel beam on the actual structure while under construction. At the time of the measurements, there was no live load and only partial dead load applied to the structure. It is believed that there must have been some three-dimensional stiffness effects, because the results of the measurements showed smaller values for the bending moments in the columns and much smaller values for the axial force in the transfer beam than those predicted theoretically and by model studies.

7.3 SOME REMARKS CONCERNING BEHAVIOR, ANALYSIS, AND DESIGN

Parametric studies for column-supported shear walls subjected to gravity loads, such as the one described above, have led to the following conclusions:

(a) The axial force in the transfer beam decreases as the column stiffness increases with respect to that of the transfer beam. As the columns become stiffer they carry more of the arch thrust directly, and thus relieve the transfer beam.

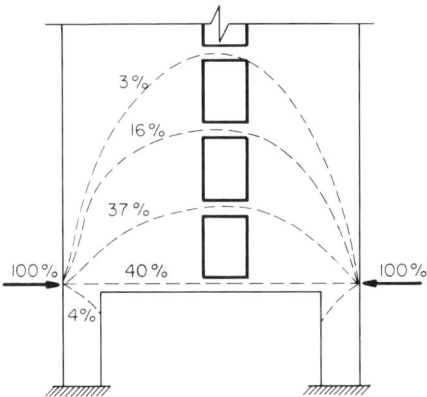

FIG. 7.12. Effect of post-tensioning.

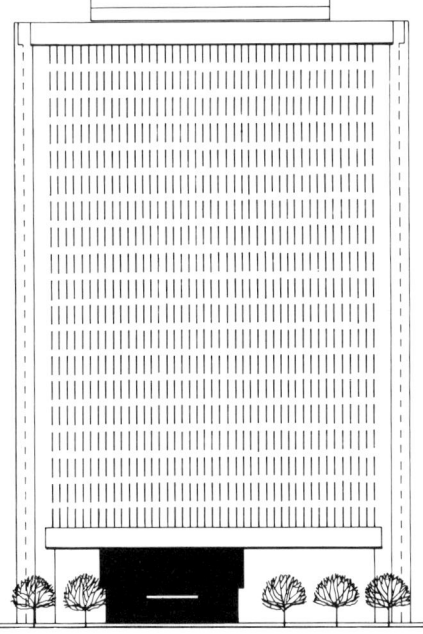

FIG. 7.13. Elevation of One Hundred Washington Square Building in Minneapolis.

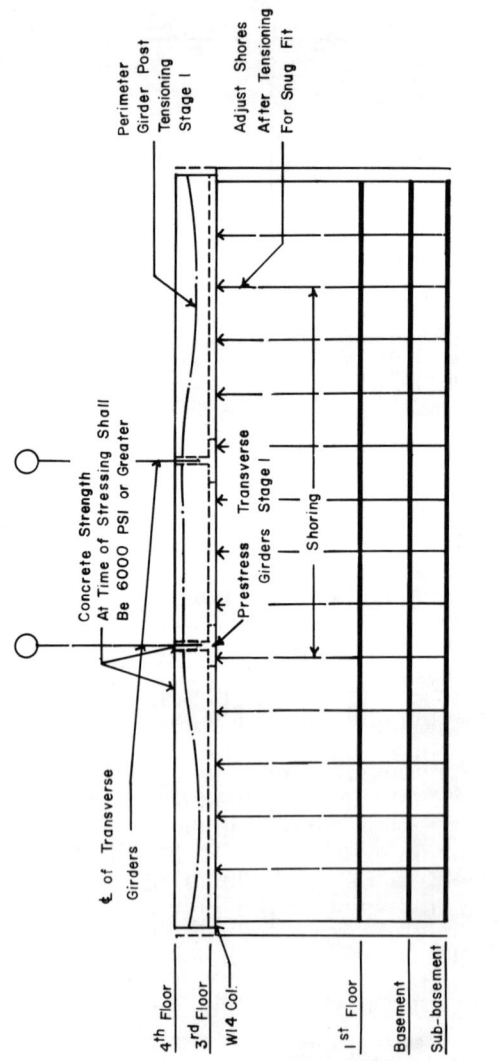

COLUMN-SUPPORTED SHEAR WALLS

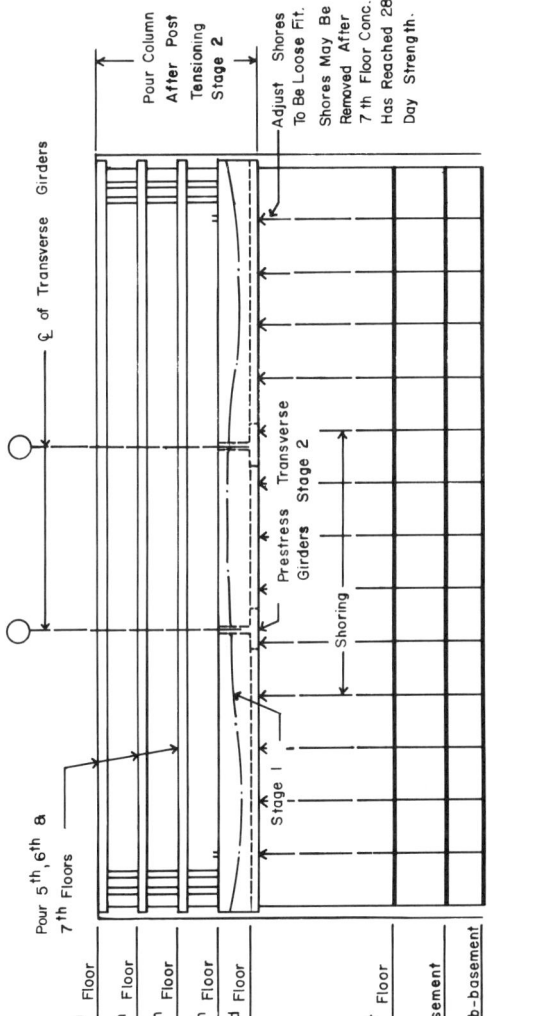

FIG. 7.14. Stage post-tensioning. Construction of north and south girders.

(b) The axial force in the transfer beam increases and the column moments decrease as the stiffness of the beam is increased with respect to that of the columns.

(c) The large axial force on the transfer beam indicates that it will probably crack at the opening. Modelling of this section of the transfer beam as a truss member (hinges at its ends) resulted in insignificant changes in the axial load in the transfer beam.

(d) In an effort to study the effect of a failure of the lowest transfer beam, this member was removed from the FEM models and the analysis rerun. The results indicated that the column moments will approximately double, quite likely resulting in a failure of the columns and a collapse of the structure. The analysis also indicated that the second transfer beam would be stressed to an axial force of about 35% that of the lowest beam under the same loading condition. Thus the lower transfer beam is a critical member and careful attention should be paid to its design, since no secondary mechanism is available to carry the loads. This conclusion does not take into account a possible three-dimensional redistribution of forces if the floor slabs are assumed to be rigid and capable of transferring the forces to other stiffer frames.

(e) If the axial force in the transfer beam is the controlling parameter in design, the designer should carefully evaluate the possibility of post-tensioning the transfer beam. Figure 7.12 shows the effect of a post-tensioning force of 100 units on a finished structure. As seen, only about 40% of the force goes into the bottom transfer beam. Thus it would seem a rather inefficient procedure to reduce the axial load in the transfer beam; on the other hand, a staged post-tensioning procedure would very likely provide much more satisfactory results. A recent example of this approach is shown in Figs. 7.13 and 7.14, in which a large transfer beam acting as the bottom chord of a multistorey Vierendeel truss was post-tensioned at three different stages of the building construction (Yee and Kim, 1983).

7.4 CONCLUDING REMARKS

The analysis and design of column-supported shear walls can be achieved by several methods. While the approximate methods and the frame idealization give conservative values for design, the finite element method seems the preferable analytical tool for this type of structure.

The designer should be aware that this is one case where being conservative is not necessarily a good design approach. The bigger the axial force the transfer beam is designed for, the larger and stiffer its section; the stiffer the transfer beam is with respect to the wall, the larger the axial force it will need to carry. If an iterative approach is used in design, a conservative approach can lead to an overdesigned transfer beam both in terms of strength and stiffness.

It is also recommended that in a structure containing shear walls supported on columns, there be a number of ordinary solid shear walls or shear walls with nominal openings which are continued to the foundations. This will provide a backup structural system to ensure the safety of the building even if a transfer beam fails.

REFERENCES

CERNY, L. (1979) Column-supported shear walls with openings. In *Planning and Design of Tall Buildings* (BEEDLE, I., Ed.), ASCE Monograph, Vol. 5, ASCE, New York, pp. 282–9.

COULL, A. and SMITH, B. S. (1967) Analysis of shear wall structures. In *Tall Buildings* (COULL, A. and SMITH, B. S., Eds.), Pergamon Press, Oxford, pp. 139–56.

ESTEVA, L. (1980) Design: General. In *Design of Earthquake Resistant Structures* (ROSENBLUETH, E., Ed.), Halstead Press, John Wiley and Sons, New York, pp. 80–3.

GREEN, D. R., MACLEOD, I. A. and GIRARDAU, R. S. (1973) Force actions in shear wall support systems. In *Response of Multistory Buildings to Lateral Forces*, SP-36, ACI, Detroit, pp. 241–56.

MACLEOD, I. A. (1967) Lateral stiffness of shear walls with openings. In *Tall Buildings* (COULL, A. and SMITH, B. S., Eds.), Pergamon Press, Oxford, pp. 223–52.

MACLEOD, I. A. (1971) *Shear Wall–Frame Interaction: A Design Aid with Commentary*, SP011.01D, Portland Cement Association, Skokie, 62 pp.

MACLEOD, I. A. and GREEN, D. R. (1973) Frame idealization for shear wall support systems. *Structural Engineer*, **51** (2), 71–4.

PAULAY, T. (1975) Design aspects of shear walls for seismic areas. *Canadian Journal of Civil Engineering*, **2**, 321–44.

SCHWAIGHOFER, J. and BARNARD, P. R. (1967) Interaction of shear walls connected solely through slabs. In *Tall Buildings* (COULL, A. and SMITH, B. S., Eds.), Pergamon Press, Oxford, pp. 157–74.

SCHWAIGHOFER, J. and MICROYS, H. F. (1969) Analysis of shear walls using standard computer programs. *ACI Journal*, **66** (Dec.), 1005–7.

SHEPERD, R. and PLUNKETT, A. W. (1983) Damage analysis of Imperial County Services Building, *ASCE Journal of Structural Engineering,* **109**(7), 1711-26.
SMITH, B. S. (1970) Modified beam method for analyzing interconnected shear walls, *ACI Journal,* **67** (Dec.), 977-80.
YEE, A. A. and KIM, C. N. (1983) One Hundred Washington Square, Structural design and construction, *PCI Journal,* **29** (1), 24-48.

Chapter 8

DESIGN OF REINFORCED CONCRETE FLAT SLABS

P. E. REGAN

Department of Civil Engineering, Polytechnic of Central London, London, UK

SUMMARY

Following a brief review of the general behaviour of flat slabs, this chapter treats two main topics — the flexibility of slab–column connections and resistance to punching. The estimation of deflections is also referred to.

NOTATION

A_{sv} Cross-sectional area of shear reinforcement
b Overall breadth of slab in an equivalent frame
b_{eff} Maximum effective slab breadth in an equivalent frame
b_e Equivalent breadth of slab, taking account of connection flexibility
c_1 Dimension of rectangular column parallel to span of slab
c_2 Dimension of rectangular column perpendicular to span of slab
D Flexural rigidity of slab
d Effective depth of slab
E Elastic modulus
e Eccentricity of load at a slab–column connection
f_c Compressive strength of concrete
f_s Stress in reinforcement
f_y Yield or 0·2% proof stress of reinforcement
h Overall thickness of slab
K Stiffness (M/θ)
k Stiffness coefficient (K/D)
l Span of slab

M Bending moment
q Uniform load — load per unit area
u Length of control perimeter used in calculations for punching
V Shear force — reaction at a column
V_s Shear force carried by shear reinforcement
v Nominal shear stress (V/ud)

α Angle between shear reinforcement and slab plane
β Angle between punching failure surface and slab plane
θ Rotation
ρ Ratio of flexural reinforcement

Subscripts
c   Column
j   Joint or connection
o   Datum value for comparison
s   Slab

8.1 INTRODUCTION

The basic form of flat slab is the simple flat plate with a solid slab supported by prismatic columns. This has great advantages in many respects but in structural terms it presents three major problems. The slab–column connections do not possess the rigidity of beam–column joints; shear is highly concentrated around the columns; and deflections tend to be large due to the minimisation of structural depth and the additive effects of bending in two directions.

Where necessary these problems can be alleviated by departures from the simple plate form of construction. Overall stability of the structure is commonly provided by shear walls and it is only in relatively low-rise buildings that slab-column frames alone have to resist horizontal loads. Column capitals (Fig. 8.1a) increase shear resistance and joint rigidity, as do drop panels (Fig. 8.1b), which also reduce deflections. Even shallow edge beams can be effective in controlling deflections and reducing stresses around external columns. Deflections throughout a slab can be reduced by prestressing or by the use of a waffle slab (Fig. 8.1c) to increase stiffness without increasing self weight.

DESIGN OF REINFORCED CONCRETE FLAT SLABS

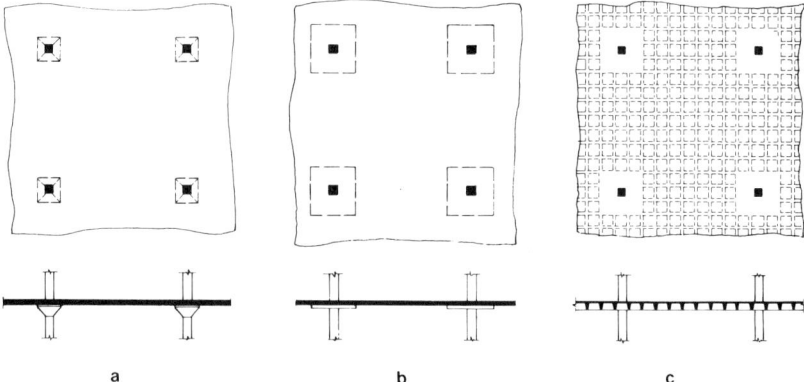

FIG. 8.1. Variations from simple flat plate construction: (a) column capitals; (b) drop panels; (c) waffle slab.

The following sections are written principally in terms of flat plates, but references to some variants are made where appropriate. Prestressed slabs are not included.

8.2 BEHAVIOUR OF FLAT PLATES

In the early years of this century there were many misconceptions about the basic equilibrium requirements of slab systems. Although the problems were soon resolved at a theoretical level, they left a legacy of incorrect ideas in codes of practice, which has only been eradicated quite recently. It is therefore appropriate to stress at the outset the fundamental requirement of equilibrium illustrated in Fig. 8.2. The figure shows a typical span in either direction subject to a uniform load q on a breadth b of the slab. Taking moments about column line (1) for the adjacent half span and denoting the shear at midspan by V:

$$M + M'_1 = \frac{qbl^2}{8} + \frac{Vl}{2} \tag{8.1}$$

Similarly for the other half span:

$$M + M'_2 = \frac{qbl^2}{8} - \frac{Vl}{2} \tag{8.2}$$

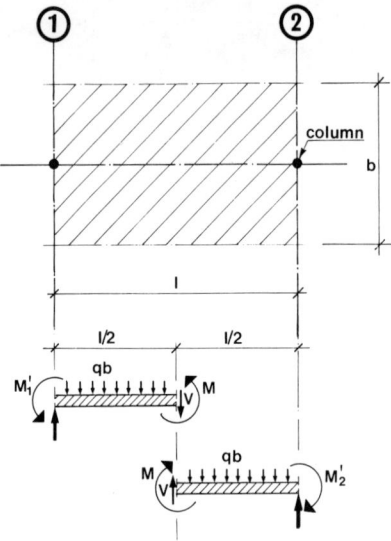

FIG. 8.2. Basic equilibrium conditions.

Adding and dividing by 2:

$$M + \tfrac{1}{2}(M'_1 + M'_2) = \frac{qbl^2}{8} \tag{8.3}$$

The requirement applies equally to spans in either direction. Thus in both directions of a slab supported at points on a rectangular grid, the sum of the midspan moment and the average of the support line moments must be equal to the free moment for the span. The free moment is slightly reduced if the finite dimensions of the columns are taken into account but this does not alter the basic fact that the full load must be carried by flexural actions in each direction.

The basic equilibrium statement is vital but does not say very much about the distribution of moments throughout a slab. A good impression of the overall behaviour of a flat plate is given by simplifying its deformations to a number of rotations between rigid segments. The boundaries between the segments can be viewed as elasto-plastic 'yield lines'. The general two-way flexure corresponds to the system of folds shown in Fig. 8.3. Superimposed on this there are local deformations due to the concentration of moments near the columns. These take two forms depending on whether the major effects in the slab are due to the vertical reactions from the columns or the moments transferred between the columns and slab.

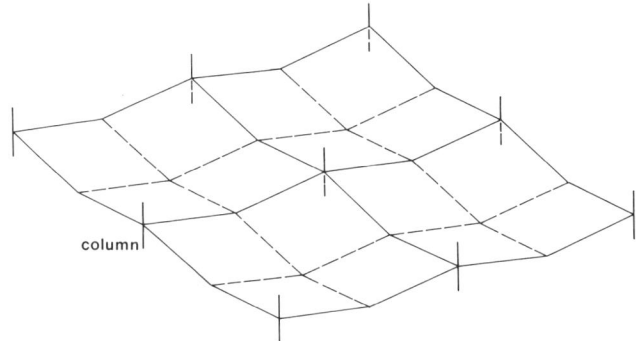

FIG. 8.3. Overall deformation of a flat plate on columns.

The vertical reactions produce the types of displacements shown in Fig. 8.4 for interior (a), edge (b) and corner (c) columns. In each case the part of the slab outside the region of local deformation moves downward while remaining horizontal. Moments transmitted from the columns to the slab cause the sorts of deformations drawn in Fig. 8.5 in which the outer parts of the slab rotate relative to the columns.

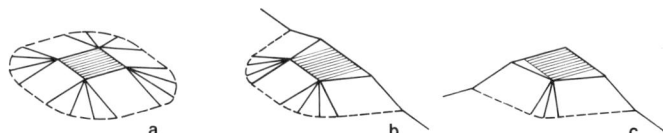

FIG. 8.4. Local vertical displacements.

At internal columns the actions of the slab are commonly more or less symmetrical around the column and the transferred moments are generally small. The deformation of Fig. 8.4(a) thus predominates over that of Fig. 8.5(a). However, at edge and corner columns, where

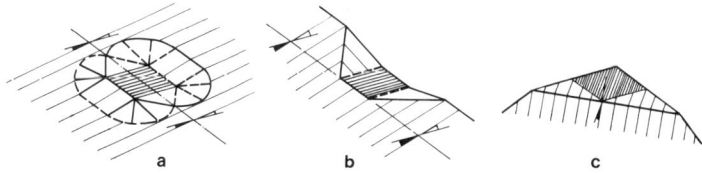

FIG. 8.5. Local rotational displacements.

moments perpendicular to the edges are not balanced within the slab, the moment transfer is much more significant and the dominant local deformations are those of Fig. 8.5(b) and (c).

Tankut (1969) tested two large slab specimens, each supported on nine columns and subjected to uniform loading in tests to failure. The crack patterns of the two models were similar and those from one of them are drawn in Fig. 8.6. It can be seen that the cracking of the bottom surface corresponds to the idealisation of Fig. 8.3, with the addition of some circumferential cracks similar to those in Fig. 8.4. For the top surface, the model of Fig. 8.3 represents the regions away from the columns, but near the columns it has to be supplemented by local deformations as in Fig. 8.4(a) for the interior column and Fig. 8.5(b and c) for the edge and corner columns.

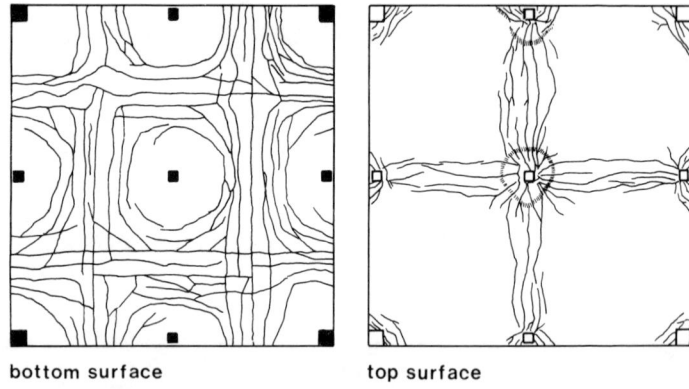

bottom surface   top surface

FIG. 8.6. Crack patterns of a slab tested by Tankut.

Since the main cracks in the bottom surface go across the full slab width, the reinforcement for positive moments due to gravity loading can be fairly uniformly distributed for all but the most elongated panel shapes. The top steel has to be concentrated toward the columns. If there is insufficient top steel near interior columns, a failure mechanism involving only the effects shown in Fig. 8.4(a) becomes possible. The mechanism is illustrated by Fig. 8.7 and does not mobilise any resistance from the reinforcement outside the circumferential positive moment yield lines. The exact dimensions of these yield lines depend upon the arrangement of reinforcement but their radius would normally be of the order of a quarter of the shorter span of the slab. A

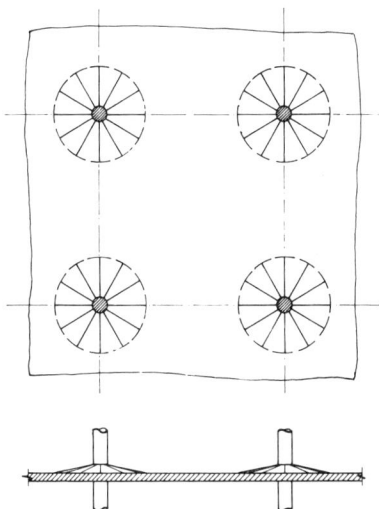

FIG. 8.7. Local failure mechanism.

large proportion of the total steel at any internal column should thus be placed within strips of widths equal to half the shorter span.

At edges and corners the extent of the cracking and eventual yield lines is much more restricted (see Figs. 8.4, 8.5 and 8.6) and is directly related to the column dimensions and not to the slab spans. Since bars not crossed by cracks remain almost unstressed, the top steel perpendicular to the edges must be highly concentrated towards the columns to be of any use and must be adequately anchored to each side of a potential yield line if it is to be fully effective.

For pure horizontal loading, only small slab zones are involved in the transfer of moments to interior as well as exterior columns and the reinforcement for the transferred moments should be highly concentrated at all slab–column connections. The concentration of moments in small slab areas causes the frame action to have a low stiffness even though there is normal continuity between consecutive slab spans.

If reinforcement is appropriately arranged and well detailed, a yield line analysis should give a reliable estimate of the ultimate resistance of a slab provided the failure is purely flexural. However, in flat slabs there is a strong interaction between the magnitudes of moments transferred to interior columns and the punching resistances of the slab–column connections. Yield line theory offers no method of assessing transferred moments and its use in design would allow, and indeed encourage, designers to take the moments to be zero for all cases of gravity loading. In reality, moments could develop, reduce the punching resistance and so lead to premature failure. To be satisfactory a general design method must include a means of estimating the moments transferred to columns.

In practice, moments are normally estimated as approximations to elastic plate solutions. There is little purpose in attempting great accuracy in such approximation as the effects of cracking of the concrete and local yielding of reinforcement lie outside the basic assumptions of elastic behaviour. If unusual accuracy is required, resort must be made to numerical methods incorporating realistic moment-curvature and torque–twist relations. Analyses can then be made by finite element methods or grillage analogy but the design process is iterative as the slab behaviour is a function of its reinforcement.

8.3 EQUIVALENT FRAME METHODS

8.3.1 Simple Equivalent Frame Method

The commonest design approach is to determine gross moments by an elastic equivalent frame analysis and to distribute them over the width of the slab following rules derived from elastic plate theory. The load bearing action in each direction is represented by that of a series of one-way spanning strips each connected to a line of columns. The stiffnesses of the frame members are determined from the gross concrete sections of the columns and the associated slab strips of breadths equal to the dimensions between panel centre lines or between a centre line and an edge. The approach respects equilibrium and gives gross moments in agreement with elastic plate analysis for elementary cases in which no moments are transferred to columns.

Even without considering moment transfer, the model can be improved somewhat by considering the load bearing in an elongated panel. As indicated in Fig. 8.8, the primary action is in the long span with moments distributed across the full breadth, but the short span moments are confined to bands close to the column lines. This concentration of moments reduces the stiffness of the short span. For a uniformly loaded slab simply supported on lines at two opposite edges the end rotations are $ql^3/24 D$. If the line supports are replaced by point supports at a spacing b, the rotations can be calculated by elastic plate theory and are increased ($\theta > ql^3/24 D$).

DESIGN OF REINFORCED CONCRETE FLAT SLABS

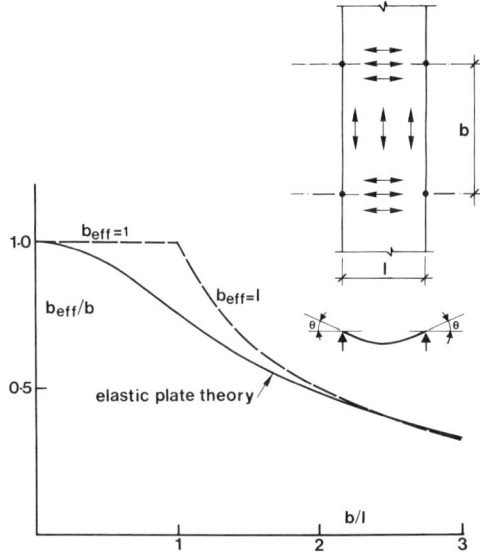

FIG. 8.8. Upper limit of effective breadth.

In relation to these rotations, which define compatibility in a frame, the actual slab on point supports can then be viewed as equivalent to one supported full width but with a reduced effective breadth (b_{eff}) such that

$$\theta = qbl^3/24\, b_{\text{eff}} D \qquad (8.4)$$

Figure 8.8 shows values of b_{eff} calculated from eqn (8.4) and plate theory solutions for θ. They can be approximated by taking b_{eff} as the lesser of b and l. This approximation is appropriate for continuous spans as well as simple ones.

8.3.2 Stiffnesses of Slab-Column Connections

The continuity between a flat plate and its columns is inferior to that of a slab-wall frame. A moment acting on a column and resisted by a plate produces a rotation greater than that which would occur if the members were connected over their full widths. Similarly, if the column provides a restraint, the deformation of the slab is greater than that for a full width connection.

Figure 8.9 illustrates a simple example representing a typical situation arising with horizontal loading. The stiffness of the assembly can be expressed by

$$\frac{M_j}{\theta} = kD \tag{8.5}$$

where k is a function of the dimensions of the column and slab.

With full width support $\theta = \theta_o$ and $k_o = 12b/l$ for a simple analysis or

$$k_o = 12\frac{b}{l}\frac{1}{(1 - c_1/l)^3} \tag{8.6}$$

if the part of the slab within the length c_1 is infinitely rigid.

The rotation of the flat plate assembly can be expressed as the sum of θ_o and an additional joint rotation θ_j giving

$$\frac{M_j}{kD} = \frac{M_j}{k_oD} + \frac{M_j}{k_jD} \tag{8.7}$$

$$\frac{1}{k_j} = \frac{1}{k} - \frac{1}{k_o} \tag{8.8}$$

where k_jD is the stiffness of the slab–column joint.

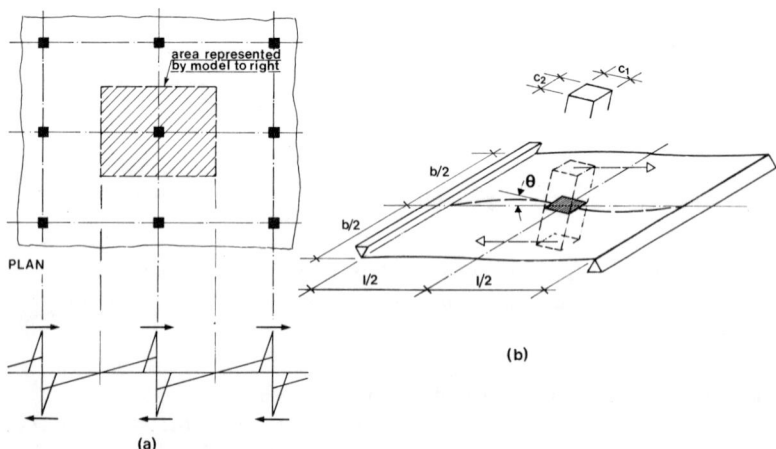

FIG. 8.9. Deformation of a slab due to a moment from a column: (a) elevation and moment diagram; (b) model of typical part of slab.

Alternatively the flat plate can be regarded as equivalent to a narrower slab of breadth b_e loaded over its full width. The equivalent breadth b_e is given by

$$\frac{b_e}{b} = \frac{k}{k_o} = \frac{k_j}{k_j + k_o} \tag{8.9}$$

Values of k or k_j can be assessed in various ways. To date theoretical proposals have been expressed in terms of the properties of gross concrete sections. The influence of cracking and eventual yielding of reinforcement have to be allowed for by empirical calibrations.

Solutions for k or k_j based on elastic plate theory fall into two principal groups depending on the assumptions made about the stiffness of the slab within the area of the column. Lower bound values are obtained if the stiffening of the slab by the column is ignored, while upper bound values correspond to the assumption that the slab is infinitely rigid within the column. Neither approach is altogether realistic. The column obviously stiffens the slab but it does not reduce the strain of the reinforcement to zero and the curvature suppressed within the column area appears as a rotation in a crack at the column face.

Figure 8.10(a), taken from the results of finite element analyses by Mehrain and Aalami (1974), shows upper and lower bound k values for square columns and square slab panels. The difference between the two increases with increasing column size. The influences of column shape

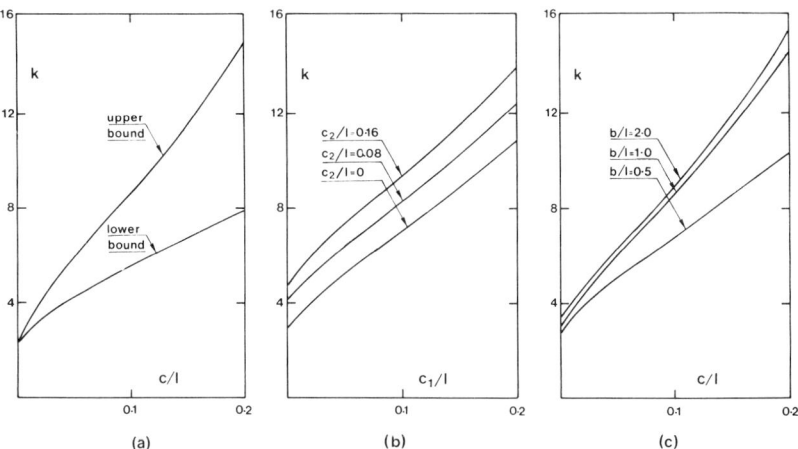

FIG. 8.10. Stiffness coefficients k from elastic plate theory. (a) $c_1 = c_2$, $b = l$, (upper and lower bound solutions from Mehrain and Aalami); (b) $b = l$, (upper bound solutions from Wong and Coull); (c) $c_1 = c_2$, (upper bound solutions from Wong and Coull.)

and panel proportions are illustrated by Fig. 8.10(b) and (c) based on upper bound solutions by Wong and Coull (1980). The column dimension parallel to the moment vector and the panel shape can be seen to have relatively little influence on k and an approximation to elastic plate solutions can be made in the form

$$k_j = A + B(c_1/l) \tag{8.10}$$

with the constants A and B theoretically dependent on the form of the elastic analysis and the expression used for k_o, i.e. eqn (8.6), the simple $k_o = 12b/l$ or an intermediate form.

In practice the assessment of the constants must be influenced by other factors related to the differences between ideal elastic structures and real reinforced concrete. It is common practice in design to base analyses of moments on stiffnesses calculated for gross concrete cross-sections with the implicit assumption that cracking degrades the stiffnesses of the various members in similar proportions and thus has little effect on distributions of moments.

This assumption is always debatable but the present case is worse than most for it. Cracking of the slab occurs in the immediate vicinity of the columns at a proportion of ultimate load considerably less than is normal for concrete members. Thus, at service loads, the deterioration of the connection stiffness is more advanced than that of the positive moment regions of the slab. Furthermore, the connection stiffness is largely dependent on torsion in the slab and where principal moments are at 45° to reinforcement the fully cracked stiffness is only half that for a similar slab with the steel parallel and perpendicular to the principal moments. Finally, the moment transfer stiffness is not defined by a unique moment-rotation relationship but depends on the vertical loading of the slab, with higher vertical loads producing more cracking and thus a lower stiffness (Regan, 1984).

If analyses for moment distributions are to be based on gross section values of D, it follows that the constants in eqn (8.10) must be reduced well below their theoretical elastic values. The fact that columns may not crack and may retain their full stiffnesses under gravity loading is not very important here as the compatibility condition at a typical connection depends essentially on the stiffnesses of the slab and the joint, with the column rotation normally being very small.

Values of k_j proposed by Regan (1981) are given in Fig. 8.11. They are relatively conservative in terms of the constant B in eqn (8.10). A physical model representing this treatment of a connection is drawn in Fig. 8.12(a).

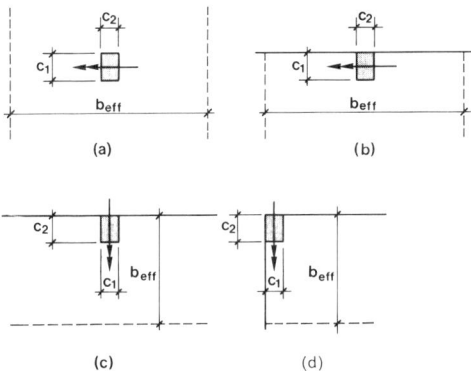

FIG. 8.11. Stiffness coefficients k_j recommended in CIRIA Report 89. (a) $k_j = 3\cdot0 + 40\,c_1/b_{\text{eff}}$; (b) $k_j = 1\cdot5 + 20\,c_1/b_{\text{eff}}$; (c) $k_j = 1\cdot5 + 20\,c_1/b_{\text{eff}}$; (d) $k_j = 0\cdot75 + 20\,c_1/b_{\text{eff}}$.

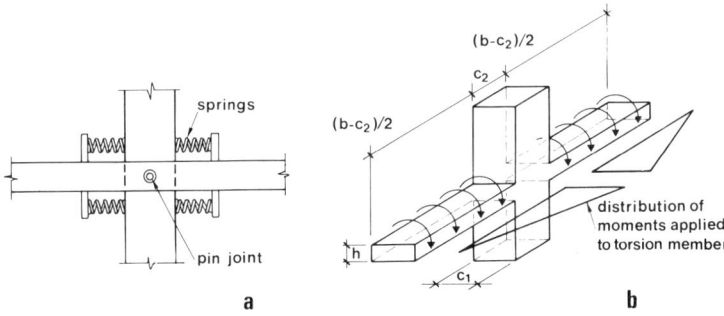

FIG. 8.12. Models of slab–column connections. (a) CIRIA 89; (b) ACI.

The ACI code (1983) does not directly use an elastic plate model to assess connection stiffnesses and instead adopts a simplified structural model. As shown in Fig. 8.12(b), the moment transfer is assumed to be effected by torsion members within the slab. These are loaded by moments from the spans distributed linearly over the width, and the rotation of the connection is taken to be the average twist of the torsion members:

$$k_j = \Sigma\, 36D \left(\frac{c_1}{b}\right) \frac{(1 - 0\cdot63h/c_1)}{(1 - c_2/b)^3} \tag{8.11}$$

with the summation being for the torsion members on either side of the column. For a connection with equal transverse spans (b), $\Sigma\, 36D$ thus becomes 72.

One feature of the ACI solution is clearly wrong — the stiffness cannot deteriorate to zero if $c_1 = 0$, as at an edge with the slab connected to a projecting column only at its inner face. The assumption of a linear distribution of load on the torsion member irrespective of the ratio b/l is also unrealistic. However, in the current state-of-the-art it is more or less meaningless to quibble about theoretical points related to elastic plates when unquantifiable errors in the influence of cracking must be of much greater significance.

The flexibility of connections influences not only the stiffnesses of members with respect to moments and rotations at the same ends but also carry-over effects. In normal circumstances the carry-over variations can be ignored.

8.3.3 Improved Equivalent Frame Methods
When joint stiffnesses or slab effective widths have been assessed it remains to incorporate them in an analysis. Here it should be recalled that the results of the connections being flexible are to reduce the restraint of the slab by the columns for vertical loading and to reduce the restraint of the columns by the slab for horizontal loading, i.e. in terms of moment distribution the member whose restraint is reduced is always that having an unbalanced moment at a joint prior to the redistribution.

Fundamentally there is no special problem in using the model of Fig. 8.12(a) and the basic compatibility condition for a node is always

$$\theta_c = \theta_s \pm \theta_j \tag{8.12}$$

A computer approach is given by Vanderbilt and Corley (1983).

In hand methods there can be some confusion due essentially to the $\pm$ of the equation. For vertical load analysis, in which the columns are treated as fixed-ended at the floors above and below the slab considered, the total column stiffness at any joint can be replaced by an equivalent value ΣK_{ec} such that

$$\frac{1}{\Sigma K_{ec}} = \frac{1}{\Sigma K_c} + \frac{1}{K_j} \tag{8.13}$$

where ΣK_c is the sum of the stiffnesses of the columns above and below the floor in question and each stiffness is $4EI/L$ where L is the storey height. For horizontal loading the commonest approach is to replace the actual slab spans by members having 'equivalent breadths' as defined by eqn (8.9). The two methods cannot be switched for the two

DESIGN OF REINFORCED CONCRETE FLAT SLABS   231

types of loading, and even for the appropriate load cases each involves some errors when end moments from the initially unloaded type of member have to be balanced.

In view of the uncertainties involved in assessing joint stiffnesses, the British code BS 8110 (1985) adopts a much simpler approach merely allowing large redistributions of joint moments M_j up to 30% at interior columns and 50% at edge and corner columns.

Irrespective of the techniques adopted, a designer has two physical choices to make in an analysis. One is whether or not to make allowance for the finite sizes of connections, i.e. in essence whether to base member stiffnesses on eqn (8.6) or the simple $k_o = 12b/l$. The question is not unique to flat slabs and in general it is adequate to use the simpler form. The ACI code (1983) includes a method of interpolation between the two limits but in practice unless member thicknesses are remarkably large in comparison with their spans its results are scarcely distinguishable from $k_o = 12b/l$.

The second choice is that of numerical stiffnesses. In calculating moment distributions it is adequate to work in terms of uncracked concrete section values and the k_j values of both the ACI code and Regan (1981) were devised with this in mind. However, if side-sway movements under horizontal loading are to be assessed realistically, eventual cracking has to be taken into account. The effects of cracking will generally be different in the columns, the joints and the negative and positive moment regions of the slab. In spite of this it is convenient and not too unrealistic to treat the slab and joints as a single member and with normal vertical loading to assess its stiffness on the basis of midspan moment conditions using an appropriate formula to interpolate between the uncracked and fully cracked rigidities. The value of D so found can be used for the slabs as a whole and for the connections. If the columns remain uncracked it is only reasonable to retain their full stiffnesses in the analysis.

Where drop panels are present, the thickness of the drops should be used in estimating the stiffnesses of the connections and the variation of slab stiffness along the spans should be taken into account. For slabs on column capitals the plan dimensions of the capitals should be used as c_1 and c_2 in calculating k_j. For equal basic rigidities (EI per unit width) a torsionless waffle slab has connection stiffnesses of the order of 2/3 those of a solid slab, but solid infill around the columns improves the connections considerably.

For all types of slab the negative design moments from the frame analysis can be reduced when designing the slab reinforcement so as to take account of the reaction within the length c_1. The use of moments at column faces is a little on the unconservative side.

8.3.4 Transverse Distributions of Moments

It has already been noted that in the short span direction of rectangular panels moments are effectively confined to widths equal to spans. Older codes which divided all panels into column and middle strips of equal widths and proportioned moments between them without regard to the span ratio were therefore unsatisfactory.

The new ACI and British codes define a column strip as having a width on either side of a column line equal to the lesser of $0.25b$ and $0.25l$. The remaining part widths ($0.25b$ or $0.5b - 0.25l$) at each side are then the middle strip. Proportions of the total moment to be resisted by each strip are tabulated irrespective of the ratios b/l. This approach is an improvement so far as short span moments are concerned but is still not very good for those in the long span, which become more uniform as b/l decreases.

Table 8.1 gives distributions of total moments calculated by elastic plate theory for slabs with small columns. The moments are divided between the bands indicated in Fig. 8.13. Table 8.1 also includes current ACI and British code proportions.

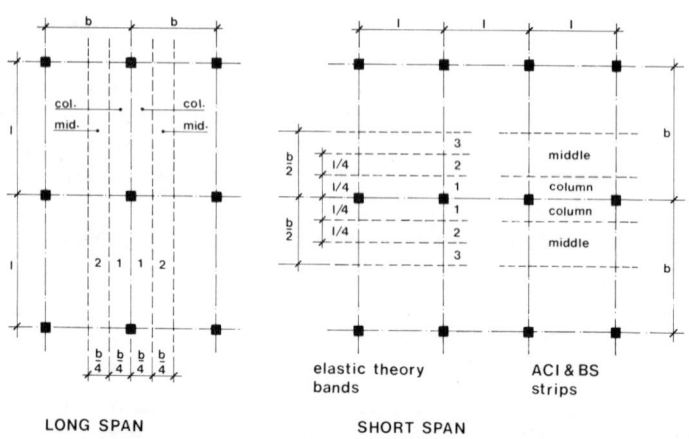

FIG. 8.13. Subdivisions of slab breadths for detailing reinforcement.

DESIGN OF REINFORCED CONCRETE FLAT SLABS

TABLE 8.1

Ratio b/l	Interior negative moments			Positive moments		
	Band 1	Band 2	Band 3	Band 1	Band 2	Band 3
0·50	0·64	0·36	—	0·51	0·49	—
0·75	0·71	0·29	—	0·55	0·45	—
1·00	0·77	0·23	—	0·61	0·39	—
1·50	0·75	0·19	0·06	0·58	0·30	0·12
2·00	0·75	0·18	0·07	0·58	0·29	0·13
ACI	0·75	0·25		0·60	0·40	
BS 8110	0·75	0·25		0·55	0·45	

If the elastic distributions are used, the reinforcement calculated for Band 3 may be less than normal minimum reinforcement. In such cases minimum reinforcement should be provided.

In the vicinity of columns the transverse distribution of negative moments is very concentrated and it is good practice to concentrate 2/3 of the column strip steel to the middle half of the width.

The above proportions are general ones determined from analyses of cases where the moments of the two spans meeting at a column are approximately equal and M_j is small. Where the transferred moment is large the concentration of negative moments at one side of the column may be increased while at the other side positive moments may be produced at the column face even though the total moment over the width b is negative. This situation needs to be taken into account in design, as failure to provide suitable local reinforcement can lead to a sudden loss of stiffness and shear strength, particularly if cracking occurs at the bottom of the slab adjacent to the column.

The ACI code (1983) recommends that the top reinforcement within a width equal to that of the column (or capital) plus $1·5h$ to each side of it should be sufficient to resist a moment

$$M = M_j \frac{1}{1 + \frac{2}{3}\sqrt{\frac{c_1 + d}{c_2 + d}}} \qquad (8.14)$$

Based on a simplified yield line analysis, Regan (1981) suggests that calculated bottom steel should be provided when $e = M_j/V > (4·7c_1 + c_2)/(\pi + 2c_2/c_1)$. It should then be designed for a moment per unit width $m = M_j/12c_1$ and be provided within the width of the column plus $1 \cdot 5h$ to each side of it.

The transverse moment distribution is a function of the cross-section of the slab. Relatively small column capitals have little effect but the influence of drop panels should be taken into account. A simple way of doing this is to distribute the reinforcement in proportion to the moments for flat plates. Due to the increased lever arm within the drops this automatically allows for the greater moments in these stiffer slab zones. The total amount of steel used is of course reduced as the moment resistance is the sum of 'steel forces × lever arms'.

In waffle slabs the reduction of torsional rigidity decreases the slab's ability to equalise moments across its width. Design solutions can be based on elastic analyses of grids with the torsional stiffness of the members taken as zero. The possibilities of redistributions of moments across the width are uncertain and the recommendations of some codes that moments can be assessed as for solid slabs are unproven.

Moment distributions can be based on the profiles developed by Beigler and Broms (1978) and shown in Fig. 8.14. The larger diagrams define the distributions of moments associated with any one column line and the smaller ones indicate the ways in which the basic figures can be combined to derive complete distributions for various panel aspect ratios.

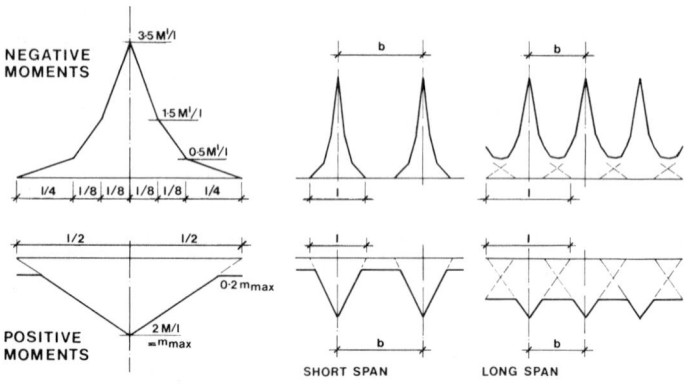

FIG. 8.14. Transverse distributions of moments in waffle slabs. After Beigler and Broms (1978). Moments can be averaged within the widths of solid infill sections. Positive moment distributions shown are for overall uniform loading.
For pattern loading, antisymmetric components give uniform moments.

DESIGN OF REINFORCED CONCRETE FLAT SLABS

At edge and corner columns, moment transfer is considerably more difficult than at interior columns. For a corner column the only active reinforcement is that crossing and effectively anchored beyond the potential yield lines of Fig. 8.15. In a floor slab the distance from the inner corner of the column to the yield line is determined by the section of maximum moment, itself a function of the distribution of shear at the column face. This distance is small. In a roof slab the yield rotation can occur in an inclined crack, thereby considerably reducing the width in which top steel can be active. Provided their diameter is not too large, column bars bent over into the slab can be a very effective form of connection reinforcement at roof level. Placing the reinforcement perpendicular to the yield line rather than parallel to the slab edges increases the stiffness of the joint but has little overall advantage as the greater moments induced tend to reduce shear resistance. Only nominal top steel is required outside the active zone.

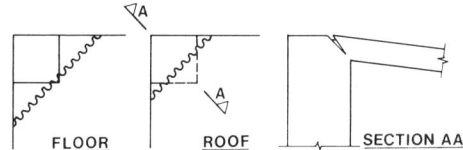

FIG. 8.15. Local yield lines at corner columns.

At edge columns, moments parallel to the slab edge are resisted by actions similar to those at interior columns but moments perpendicular to the edge produce cracking of the type shown in Fig. 8.16(a). For pure moment loading the idealised failure mechanism is shown in Fig. 8.16(b).

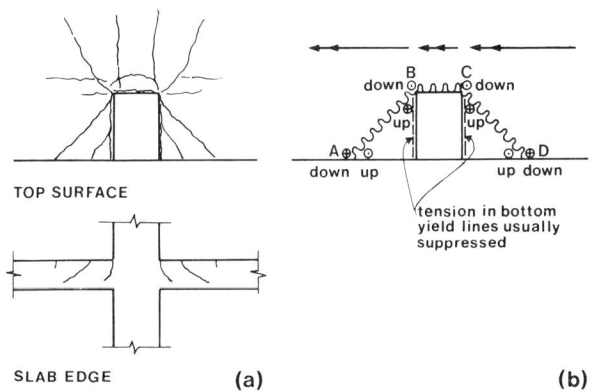

FIG. 8.16. Cracking, local yield lines and nodal forces at edge columns.

The presence of a shear load generally suppresses the tension-in-the-bottom lines at the column sides and the commonest failure mechanism is a complex one involving rotations at inclined cracks. The simple model is, however, an adequate basis for the calculation of local flexural resistance provided the moments due to the bottom bars are neglected.

When the ordinary yield line theory is applied, half of the torsional resistance at each side of the column is provided by the nodal forces indicated in Fig. 8.16(b), or in effect by torsion at the slab edge. If the reinforcement at the edge of the slab is sufficiently closely spaced and has enough of a straight vertical portion to act as torsion reinforcement, the system of Fig. 8.16(b) should be realisable. With normal spacings and the almost continuously curved shapes of U-bars in relatively shallow slabs it seems that edges should be regarded as being without torsion reinforcement. In this case the useful top steel is limited to that crossing the line ABCD in Fig. 8.16(b) and effectively anchored beyond it.

Most codes of practice offer little specific guidance on the detailing of reinforcement at exterior columns. The new British code is an exception and specifically limits the widths of the slab to be assumed active in moment transfer to those indicated in Fig. 8.17. The limitations apply to both the arrangement of reinforcement and the upper limit of moment resistance determined by compression of the concrete. The upper limit moments are relatively small and a redistribution of the moments obtained from an equivalent frame analysis is permitted with a reduction of M_j by up to 50%.

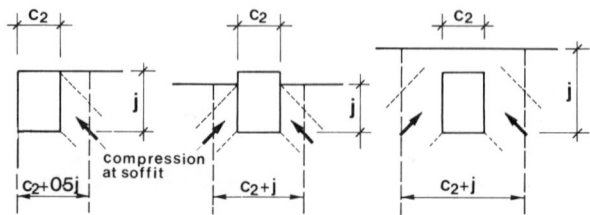

FIG. 8.17. Breadths for reinforcement at corner and edge columns (BS 8110).

The widths suggested in the British code appear appropriate so far as the upper limits are concerned since the horizontal compression in the underside of the slab must reach the column by paths such as those in Fig. 8.17. The widths may appear rather conservative in terms of the distribution of the reinforcement. However, the yield lines at the sides of the column could generally be taken to be at 45° to the slab edge and the width within which steel should be effective is then determined by the possibility of anchoring it satisfactorily. This clearly leads to a width less than j on either side and $j/2$ in Fig. 8.17 may be cautious but not unreasonable. If a greater width has to be used it is necessary to detail reinforcement at the slab edge to resist a torsion equal to the total moment coming from the bars outside the breadth above.

In considering the rather limited width within which reinforcement can be deployed at both edge and corner columns it should be remembered that the moment to be resisted is that at the inner corner or side of the column, i.e. $M_j - Vc_1/2$. The term $Vc_1/2$ can be very significant. It is not clear in BS 8110 whether the 50% permissible reduction applies to M_j or $M_j - Vc_1/2$, but the latter would seem more reasonable.

8.4 PUNCHING SHEAR

8.4.1 Symmetric Interior Punching

The case of punching most amenable to analysis is that of a symmetric failure around an interior column and this is the only case treated fully by most theories. A review of theoretical work on punching is beyond the scope of this chapter and interested readers are referred to recent state-of-art reports by ASCE-ACI Committee 426 (1974) and Regan and Braestrup (1985).

The best known theoretical model is that due to Kinnunen (1963) and Kinnunen and Nylander (1960) which treats the region of slab bounded by a line of radial contraflexure and assumes rigid body rotations of the segments between radial cracks as indicated in Fig. 8.18. The strains of the reinforcement and the tangential deformation of the concrete can be calculated as functions of the slab rotation and thence the load. The load is assumed to be transferred to the column primarily by compression in an imaginary conical shell with a geometry determined by the position of the neutral plane. The shell is assumed to be able to resist very high compressive stresses due to the triaxial nature of the real stress state, and the criterion of failure adopted is a limit on the tangential strain on the compressed face below the root of the shear crack. It is argued that excessive tangential deformations initiate delamination of the compression zone and the shell.

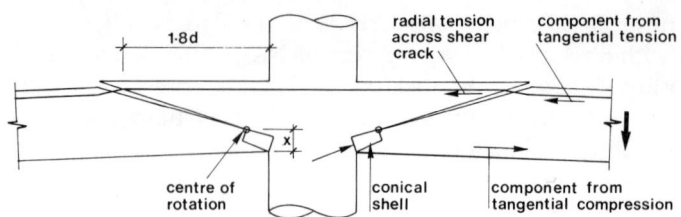

FIG. 8.18. The Kinnunen/Nylander punching model.

The other major theoretical approach is that of upper bound rigid plasticity developed by Braestrup *et al.* (1976) and others. Failure is assumed to occur with a vertical separation at the inclined failure surface and the work done in the process of a unit separation at any element of the surface is expressed in terms of the yield or failure criterion for the concrete and the local angle between the surface and the direction of motion. The shape of the failure surface and thence the ultimate load are calculated by minimising the total internal work.

These two theories differ considerably in many respects. For example, in Kinnunen's model the ratio of flexural reinforcement is a major parameter determining V_u, while in Braestrup's theory it has no influence.

Faced with the lack of a generally accepted theoretical approach most codes of practice have adopted rather simplistic solutions expressed in terms of nominal shear stresses (V/ud) on vertical sections through the slab at specified perimeters. The length u of the control perimeter depends on its distance from the column, and, since all codes have been calibrated against much the same test data, those specifying greater distances adopt lower limit stresses.

Many recommendations such as those of the ACI (1983) and CEB-FIP (1978) use perimeters at distances $d/2$ from columns together with shear stresses much greater than those for one-way slabs. If this approach is applied without correction to large columns, punching strengths can be considerably overestimated. The CEB-FIP code makes a correction by allowing the high stresses only over limited lengths near the column corners and using the basic one-way values for the remainder of the perimeter. The technique is numerically successful but its application at circular columns is highly artificial.

The British code BS 8110 (1985) was derived on the basis that the stresses for punching should be the same as those for one-way shear and it specifies a perimeter $1.5d$ from a column. This avoids the need for special treatments of large supports.

DESIGN OF REINFORCED CONCRETE FLAT SLABS

The shapes of control perimeters vary from code to code as shown by Fig. 8.19. For equal lengths of column perimeters ACI 318-83 allows more load at a square than a circular column, which is incorrect. The CEB–FIP code treats the two shapes equally, while BS 8110 gives an advantage to the circular column. This agrees with test results, but the British code seems to have problems with some other column shapes such as the 'L' in Fig. 8.19.

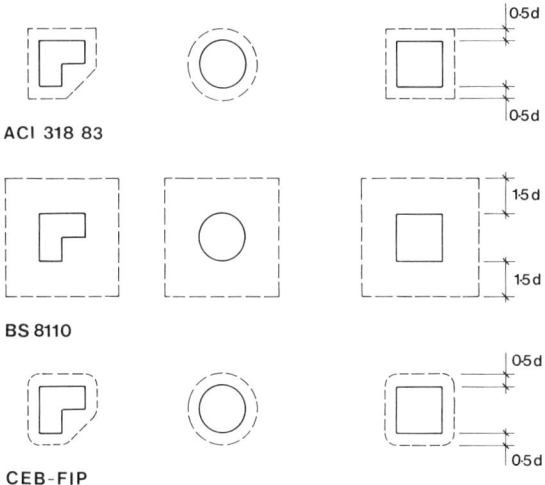

FIG. 8.19. Code of practice control perimeters at interior columns.

The parameters taken to affect the limiting shear stress also vary. They include the concrete compressive strength (f_c), the ratio of flexural reinforcement (ρ) and the effective depth of the slab (d). Their treatments in three codes are summarised in Table 8.2 but with various limits on maximum values for calculation purposes omitted.

TABLE 8.2

Code	Influences on v of				V_{test}/V_{calc}	
	f_c (N/mm²)	ρ (%)	d (m)	f_y (N/mm²)	Mean	Coefficient of variation
ACI 318-83	$\sqrt{f_c}$	None	None	None	1·46	0·16
BS 8110	$\sqrt[3]{f_c}$	$\sqrt[3]{\rho}$	$1/\sqrt[4]{d}$	None	1·06	0·09
CEB–FIP	$f_c^{2/3}$	$1 + 50\rho$	$1·6 - d$	None	1·53	0·16

A survey of test data is included in Regan and Braestrup's (1985) report. For 68 slabs with thicknesses ≥ 100 mm, the means and coefficients of variation of the ratios of experimental ultimate strengths to calculated characteristic strengths are as in Table 8.2. (Note: the figures given here for BS 8110 are correct; those in the Regan/Braestrup report are based on an earlier draft of the code.)

8.4.2 Shear Reinforcement

Shear reinforcement can increase both the punching strength and ductility of a slab–column connection. Various types are shown in Fig. 8.20. The first three act in a conventional way but shear heads are a form of composite construction. Their design is adequately covered by ACI 318-83 (1983) and they are not treated further here.

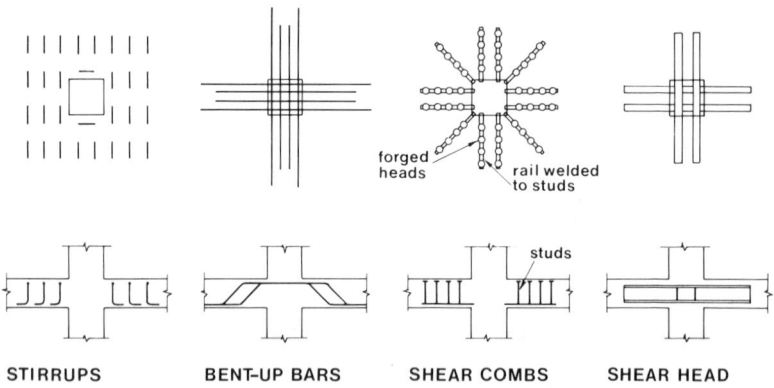

FIG. 8.20. Examples of shear reinforcement around columns.

Different modes of punching failure possible at connections with shear reinforcement are shown in Fig. 8.21. Failure outside the reinforced zone is the most common in tests, and resistance to it can be calculated by normal punching equations if a perimeter just inside the bottom anchorages of the reinforcement is substituted for the column perimeter. The perimeter can be defined by assuming the failure surface to be inclined at $45°$ in the cover (see Fig. 8.21).

Failure between the column and the innermost shear reinforcement can be avoided if the inner steel is sufficiently close ($0 \cdot 3$–$0 \cdot 5d$) to the column but, if a calculation is required, the resistance at a surface with an inclination β can be estimated as $V_o \tan \beta \geq V_o$, where V_o is the resistance of an otherwise similar slab without shear steel failing at a surface reaching to the column face.

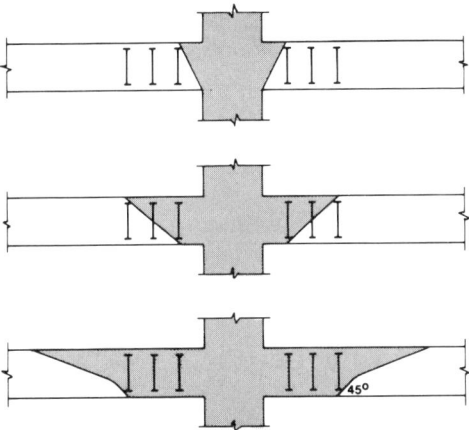

FIG. 8.21. Modes of punching failure in slabs with shear reinforcement.

Resistance to failure at a surface crossed by shear reinforcement is a combination of components from the steel and concrete. The concrete component increases with β but the rise is not significant for $\beta < 45°$. Thus a simple and satisfactory approach is to consider only 45° surfaces and to take

$$V = CV_o + V_s \qquad (8.15)$$

where V_o is the strength of an otherwise similar slab without shear reinforcement in the area in question failing at a surface reaching the soffit at the same perimeter as the surface considered; C is a coefficient taking account of the reduction of the concrete component due to the deformations required to stress the reinforcement; and $V_s = \Sigma A_{sv} f_s \sin \alpha$ is the sum of the vertical components of the forces in the shear reinforcement crossing the surface in question and adequately anchored to both sides of it.

In general C decreases as f_s increases. For properly detailed shear reinforcement, conditions at failure can be taken to correspond to $f_s = f_y$ and $C = 0.75$. BS 8110 seems very hopeful with its $C = 1.0$, permitted spacings of shear steel equal to $0.75d$ and $1.5d$ in the radial and tangential directions and V_s calculated for two layers $0.75d$ apart.

At interior connections with eccentric loading and at exterior connections, shear reinforcement should be designed for the maximum shear, and the same steel should be provided all around the perimeter.

8.4.3 Interior Slab–Column Connections with Eccentric Loads

Where an unbalanced moment is transmitted from a slab to a column, irrespective of the reinforcement, the shear effects are increased and the design shear stress due to a vertical load V and a transferred moment $M_j = Ve_x$ is generally expressed as

$$v = \frac{V}{ud}\left(1 + C\frac{e_x}{L_x}\right) \tag{8.16}$$

in which C is a constant and L_x is a dimension in the plane of the slab. If there are no nearby boundaries, L_x is a function of the column dimensions and slab thickness.

Values of C and L_x can be estimated in various ways. Codes of practice have adopted relatively simple models or equations. ACI 318-83 gives a formula for the proportion of the moment to be resisted by uneven shear and assumes a linear distribution of shear around the perimeter used for punching calculations. Thence

$$\frac{C}{L_x} = \frac{6(c_x + c_y + 2d)}{\left[1 + \frac{3}{2}\sqrt{\frac{c_y + d}{c_x + d}}\right]\left[(c_x + d)^2 + d^2 + 3(c_x + d)(c_y + d)\right]} \tag{8.17}$$

The British and European codes take:

BS 8110 $\qquad \dfrac{C}{L_x} = \dfrac{1\cdot 5}{(c_x + 3d)}$ \hfill (8.18)

CEB–FIP $\qquad \dfrac{C}{L_x} = \dfrac{1\cdot 5}{\sqrt{(c_x + d)(c_y + d)}}$ \hfill (8.19)

Figure 8.22 compares these codes. Available test data are reviewed by Regan and Braestrup (1985). Few examples of especially large or small columns are included and very few for $(c_y + d)$ significantly less than $(c_x + d)$, but within these limitations the test data support the ACI and British methods rather than the CEB–FIP code.

DESIGN OF REINFORCED CONCRETE FLAT SLABS

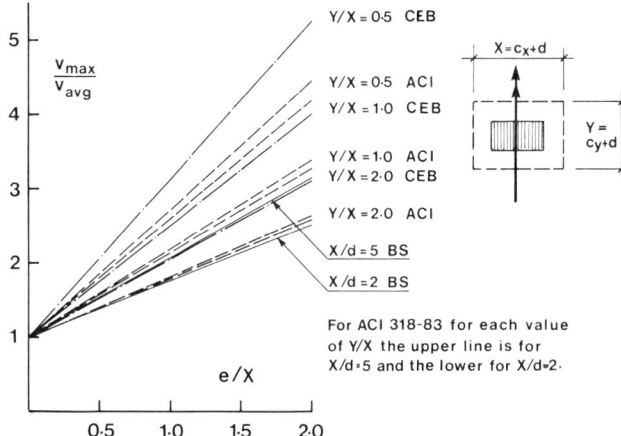

FIG. 8.22. Influence of load eccentricity on maximum shears at interior columns.

8.4.4 Connections with Edge Columns

Control perimeters at edge columns can be constructed following the same rules as for interior connections. Where a column is near but not at an edge, calculations should be made for both a three-sided edge perimeter and a four-sided interior one with the effects of eccentricity accounted for in both cases.

So far as the three-sided perimeter is concerned eccentricity parallel to the free edge has an effect similar to that at an internal column and eqn (8.16) remains applicable. The influence of a moment in the perpendicular direction is more debatable. Eccentricity is commonly defined with respect to the centroid of the control perimeter, but this may well credit a very nominal device with an unwarranted physical significance. It is, however, the approach of both the ACI and CEB-FIP codes. ACI 318-83 adopts a linear distribution of shear, similar in principle to its treatment of internal connections, and obtains a sharp moment–shear interaction. The CEB uses the same numerical equation as for interior columns and is thus less severe.

Either of these methods implicitly assumes that the moment creates uplift at the slab edge and thus that the shear at the inner column face is greater than the total applied load. One result is that the shear resistance at a column with three sides in contact with a slab can be predicted to be less than that at a column joined to the slab only at its inner face. From common sense this seems unlikely and this view is supported by available test results.

If the slab edge is not specially reinforced for torsion, the torsion cracking which occurs at relatively early loads must practically remove any uplift. The concentration of the full load to the inner face alone is then the worst conceivable effect of moment–shear interaction and the residual shear strength should correspond to the reduced control perimeter of Fig. 8.23.

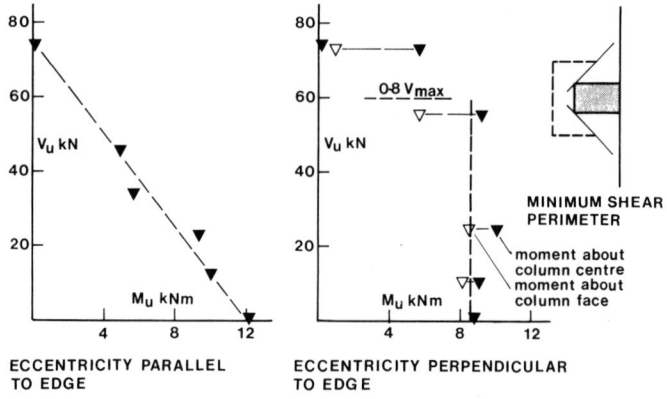

FIG. 8.23. Ultimate moment–shear interactions at edge columns.

In the British code a simple 25% allowance is made for the variation of shear around the control perimeter caused by a moment perpendicular to the slab edge. Combining this with the effect of eccentricity parallel to the edge

$$v_{max} = \frac{V}{ud}\left(1\cdot 25 + \frac{1\cdot 5e}{c + 3d}\right) \qquad (8.20)$$

where e and c are in the direction parallel to the edge.

For normal column shapes this seems reasonable, and comparisons with test results such as those by Stamenkovic and Chapman (1972) shown in Fig. 8.23 are satisfactory. It may be noted from the figure that moment and not shear failures govern for large eccentricities perpendicular to the edge. Equation (8.20) might be unsafe where the ratio of the column dimensions perpendicular and parallel to the slab edge is large, as the constant 1·25 might be too small to account for the concentration of shear to the inner face. In such cases the reduced perimeter of

DESIGN OF REINFORCED CONCRETE FLAT SLABS   245

Fig. 8.23 could be used directly.

Where the slab edge is reinforced for torsion the situation probably depends on the detailing of the reinforcement.

The above paragraphs have been concerned with inward eccentricity. Where the eccentricity is outward, conditions are worse and the only simple and safe approach for the present appears to be to design the connection as one between the column and two edge beams.

8.4.5 Connections with Corner Columns

With eccentricity of load toward the interior of the slab the maximum shears at a corner column connection are in the vicinity of the column's inner corner. The behaviour can be thought of in terms of punching by considering perimeter (a) of Fig. 8.24 or as one-way shear in a wide diagonal beam corresponding to perimeter (b). In either case an allowance should be made for the peak shear near the corner of the column being higher than the average for the perimeter. However, the allowance cannot be considered in isolation from the distance of the control perimeter from the column.

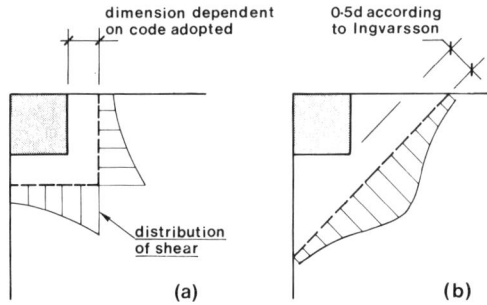

FIG. 8.24. Control perimeters for shear at corner columns.

Ingvarsson (1977) recommends a diagonal perimeter at a distance $d/2$ from the column corner. For bending at 45° to the slab edges he makes no direct allowance for non-uniformity of shear and takes

$$v_{max} = \frac{V}{d\,[d + \sqrt{2}\,(c_1 + c_2)]} \tag{8.21}$$

Some allowance for non-uniformity is made for unsymmetrical bending.

BS 8110 uses a punching perimeter at $1\cdot 5d$ from the column and having a square corner. A 25% allowance is made for shear variation around the perimeter and

$$v_{max} = \frac{1 \cdot 25V}{d(3d + c_1 + c_2)} \tag{8.22}$$

For normal column sizes the differences between the two equations are insignificant and test results show either to be acceptable. For large columns ($c > 3d$) there is no experimental data available but it may be anticipated that either equation could become unsafe as neither incorporates any term relating the concentration of shear to the ratio c/d.

8.5 DEFLECTIONS

In considering serviceability in relation to deflections it must be realised that mid-panel displacements are generally significantly greater than those mid-way between columns. The designer's first task must then be to decide upon the criteria of acceptability of deflections in the structure in question.

Once the criteria are decided, reasonable estimates of deflections due to vertical loading can be obtained from the moments found in the frame analysis if the slab stiffness is assessed with due regard to cracking and tension stiffening. The calculations can be based on average moments and stiffnesses for the effective slab breadths (b_{eff}). The refinement of treating column and middle strips separately is not justified and does not normally improve accuracy. The flexural tensile strength of the concrete should be assessed conservatively to allow for the actual non-uniformity of moments and a value of $0 \cdot 5 \sqrt{}$cylinder strength seems appropriate. Deflections mid-way along column lines are calculated for the frame in the appropriate direction, while mid-panel deflections can safely be taken as the sums of these for two orthogonal frames. The results are good for edge panels but deflections of interior panels tend to be overestimated.

8.6 CONCLUDING REMARKS

Although flat slab floors have some structural disadvantages they are likely to remain popular and can meet a very wide range of demands if sensible use is made of such possibilities as drop panels, waffle forms and prestressing.

Current treatments of flexure are relatively rational, and serious uncertainties arise only in regard to horizontal loading. In most cases these can be avoided by the use of shear or core walls, although in tall buildings some difficulties do remain in apportioning forces between the cantilever shear walls and the flat slab frames. Treatments of punching shear are not rational but adequate empirical equations are now available for most normal circumstances.

One factor ignored in this chapter is compressive membrane action, which is sometimes claimed to offer very considerable strength enhancements. While membrane or shallow dome effects can greatly increase resistance to a local load, the evidence of tests of multi-panel structures under overall loading is that it has little effect in normal design situations.

REFERENCES

ACI 318-83 (1983) *Building Code Requirements for Reinforced Concrete,* American Concrete Institute, Detroit.

ACI 318R-83 (1983) *Commentary on Building Code Requirements for Reinforced Concrete,* American Concrete Institute, Detroit.

ASCE-ACI Task Committee 426 (1974) Shear strength of reinforced concrete members — slabs. *Journal of the Structural Division, ASCE,* **100** (ST8), 1543–91.

BEIGLER, S.-E. and BROMS, C.-E. (1978) *Dimensioneringsmall för Kupolbjälklag,* Rapport R85: 1978, Statens Råd för Byggnadsforskning, Stockholm.

BRAESTRUP, M. W., NIELSEN, M. P., JENSEN, B. C. and BACH, F., *Axisymmetric Punching of Plain and Reinforced Concrete,* Report R75, Structural Research Laboratory, Technical University of Denmark, Lyngby.

BS 8110 (1985) *The Structural Use of Concrete,* British Standards Institution, London.

CEB-FIP (1978) *Model Code for Concrete Structures* (English translation), Bulletin D'Information N125-E, Comité Euro-International du Béton, Paris.

INGVARSSON, H. (1977) *Betongplattors hållfasthet och armerings utformning vid hörnpelare,* Meddelande Nr 122, Institutionen för Byggnadsstatik, Kungliga Tekniska Högskolan, Stockholm.

KINNUNEN, S. (1963) Punching of concrete slabs with two-way reinforcement. *Transactions of the Royal Institute of Technology,* No. 198, Stockholm.

KINNUNEN, S. and NYLANDER, H. (1960) Punching of concrete slabs without shear reinforcement. *Transactions of the Royal Institute of Technology,* No. 158, Stockholm.

MEHRAIN, M. and AALAMI, B. (1974) Rotational stiffnesses of concrete slabs. *ACI Journal,* **71** (9), 429–35.

REGAN, P. E. (1981) *Behaviour of Reinforced Concrete Flat Slabs,* CIRIA Report 89, Construction Industry Research and Information Association, London.

REGAN, P. E. (1984) Discussion of 'Moment-rotation characteristics of flat plate and column systems' by J F Mulcahy and J M Rotter, *ACI Journal,* **81** (1), 97.

REGAN, P. E. and BRAESTRUP, M. W. (1985) *Punching Shear in Reinforced Concrete — A State of the Art Report,* Comité Euro-International du Béton, Lausanne.

STAMENKOVIC, A. and CHAPMAN, J. C. (1972) *Local Strength of Flat Slabs at Column Heads,* CIRIA Report 39, Construction Industry Research and Information Association, London.

TANKUT, A. T. (1969) The behaviour of reinforced concrete flat plate structures subjected to various combinations of vertical and horizontal loads. PhD Thesis, Imperial College, London.

VANDERBILT, M. D. and CORLEY, W. G. (1983) Frame analysis of concrete buildings. *Concrete International,* **5** (12), 33–43.

WONG, Y. C. and COULL, A. (1980) Effective slab stiffness in flat plate structures. *Proceedings Institution of Civil Engineers, London, Part 2,* **69** (September), 721–35.

Chapter 9

PROGRESSIVE COLLAPSE OF SLAB STRUCTURES

DENIS MITCHELL and WILLIAM D. COOK
Department of Civil Engineering and Applied Mechanics, McGill University, Montreal, Canada

SUMMARY

In this chapter some historical background is given on progressive collapse of precast concrete structures and cast-in-place slab structures. The initiation of failure and the failure progression of these structures which are susceptible to progressive collapse are described. A general philosophy of providing structural integrity in the form of alternate load paths is discussed. Theoretical models for predicting the tensile membrane response of two-way slabs are compared and behavioural observations are made from some of the key experimental studies on slabs. Recent developments towards a design method for providing structural integrity in two-way slab structures are also presented.

NOTATION

A_{sb} Minimum area of effectively continuous bottom bars in direction l_n anchored into reaction area of support
f_{ps} Stress in prestressed reinforcement
f_y Specified yield strength of non-prestressed reinforcement
l_d Tension development length
l_n Clear span
l_x, l_y Clear span in the short and long directions respectively
l_2 Distance measured from the centreline of the panel on one side of the one-way catenary to the centreline of the panel on the other side of the catenary

M	Moment transferred to column by slab
M_0	Pure moment capacity
T_c	Force in the reinforcement in the one-way catenary
T_x, T_y	Force per unit length in the reinforcement in the x and y directions respectively
t	Slab thickness
V	Shear transferred to column by slab
V_0	Pure shear capacity
w	Uniformly distributed load per unit area of the slab
w_s	Unfactored service loading per unit area of the slab, but not less than twice the self-weight of the slab
δ	Central deflection of the slab
ε	Strain in the reinforcement of the one-way catenary
$\varepsilon_x, \varepsilon_y$	Strain in the reinforcement in the x and y directions respectively
ρ_x, ρ_y	Ratio of area of steel reinforcement to area of concrete in x and y directions respectively
ϕ_s	Material resistance factor for reinforcing steel
ϕ_p	Material resistance factor for prestressing steel

9.1 INTRODUCTION

9.1.1 Definition of Progressive Collapse

The 1968 Ronan Point collapse, which resulted in a public inquiry as reported by Griffiths *et al.* (1968), has been instrumental in making engineers aware of the 'possibility of a chain reaction or progressive collapse'. A localized failure due to a gas explosion in a corner apartment on the eighteenth floor of this twenty-two storey precast concrete apartment building led to a chain reaction collapse of the entire corner of the structure. A view of the structure after collapse is given in Fig. 9.1. Such a widespread propagation of an initial local failure has been termed 'progressive collapse'. Structures should not behave as a 'house of cards' and should not display a chain reaction failure mode commonly referred to as the 'domino effect'. Instead structures should display structural integrity in the event of a local failure caused by abnormal loading.

9.1.2 Abnormal Loadings

'Abnormal loading' is a term used to describe the various causes of initial failures in structures. Abnormal loads are loads which are not normally considered in design and are generally those loads which are considered to have a low probability of occurrence (Burnett *et al.*, 1973; Burnett, 1975). These include

(a) pressure loading due to gas explosions, boiler failures, bombs, sonic boom, etc.
(b) impact caused by vehicles, aeroplanes, construction cranes, falling debris, etc.
(c) construction errors, unauthorized alterations, improper maintenance, etc.
(d) natural hazards such as fires, floods, tornadoes, etc.

FIG. 9.1. Ronan Point collapse, 1968. (Photograph courtesy of B. Stafford Smith.)

It is obvious that it is not practical to design all the elements of a structure to withstand all of the possible abnormal loads.

9.1.3 Philosophy of Preventing Progressive Collapse

The two basic philosophies in designing to prevent progressive collapse are

(a) design to resist abnormal loading (i.e. 'prevent local failure'); and
(b) provide alternate load paths (i.e. 'carry load around local failure').

As a result of the Ronan Point collapse, many codes required a specific design to prevent progressive collapse by either of these two methods. For example, in Great Britain the Fifth Amendment to the building regulations (Ministry of Housing and Local Government, 1970) requires that either of the two following requirements must be satisfied:

(a) design each load bearing element for a specified pressure of 5 psi (34 kPa) in lieu of determining the loading from a gas explosion; or
(b) based on the notional removal of individual load bearing elements provide alternate load paths in order to limit the region of damage.

'Specified abnormal' loading is not only a contradiction of terms but can lead to uneconomical designs and designs which may not prevent progressive collapse. The specified loading may be extremely severe (as in the case of the 5 psi pressure loading) but will not cover all cases of abnormal events and therefore may not prevent a local failure from occurring. Provision of alternate loading paths can result in economical designs and can prevent local damage from spreading. This method of design which acknowledges the inevitability of a local failure provides a more sensible approach to this complex problem.

9.2 PRECAST CONCRETE STRUCTURES — A BRIEF OVERVIEW

9.2.1 Failure Initiation and Progression in Precast Concrete Buildings

The report on the Ronan Point collapse by Griffiths *et al.* (1968) focused on the lack of redundancy and the lack of 'alternate paths to resist the loads previously borne by the failed parts'. Figure 9.2 illustrates a possible sequence of events after the failure of a wall panel in a precast concrete building which has not been properly designed to resist progressive collapse. In such a precast concrete wall panel structure a significant abnormal loading such as a gas explosion usually results in the loss of the entire load carrying element due to the lack of continuity and lack of reinforcement across the joints. If the structure has not been designed to carry the load around such failed members by means of alternate load paths, then a progression of failure will result (see Fig. 9.2). In this example the failure of one element leads to the loss of support of the elements above the failure region (see Fig. 9.2b), resulting in progression of failure due to the debris loading (Fig. 9.2c).

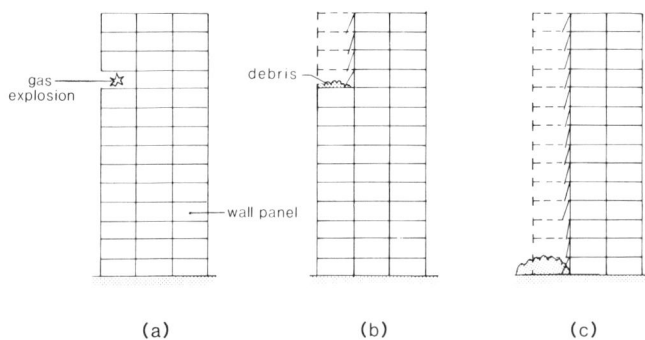

FIG. 9.2. Sequence of failures in a precast concrete building not designed to prevent progressive collapse. (a) Initial failure; (b) failure progression due to loss of support; (c) failure progression due to debris loading.

Designers are now fully aware that for these structures particular care needs to be taken in order that they do not behave as a 'house of cards'. The trend towards the use of large panel construction in North America makes this problem even more acute.

9.2.2 Current Trends in Providing Structural Integrity of Precast Wall Panel Structures

Figure 9.3 illustrates the manner in which structural integrity, in the form of alternate load paths, may be provided in precast concrete wall panel structures (Portland Cement Association, 1976; 1977). The essential feature of this approach is to provide continuous reinforcement across joints of elements to enable tie forces to develop and thus provide an alternate load path. Figure 9.3(a) shows the resisting mechanism after the loss of a load bearing element. It is crucial that this mechanism be capable of supporting the elements above the failure region as shown. Equilibrium together with a simple compatibility relationship (i.e. assuming that the strains in the horizontal ties vary from zero at the base of the resisting mechanism to the tensile rupture strain at the top of this region) provide a simple understandable tool for designing the horizontal ties. The bottom quarter of each vertical joint in each storey is filled with drypack in order to be able to transfer the compression (see Fig. 9.3a) (Portland Cement Association, 1977). The

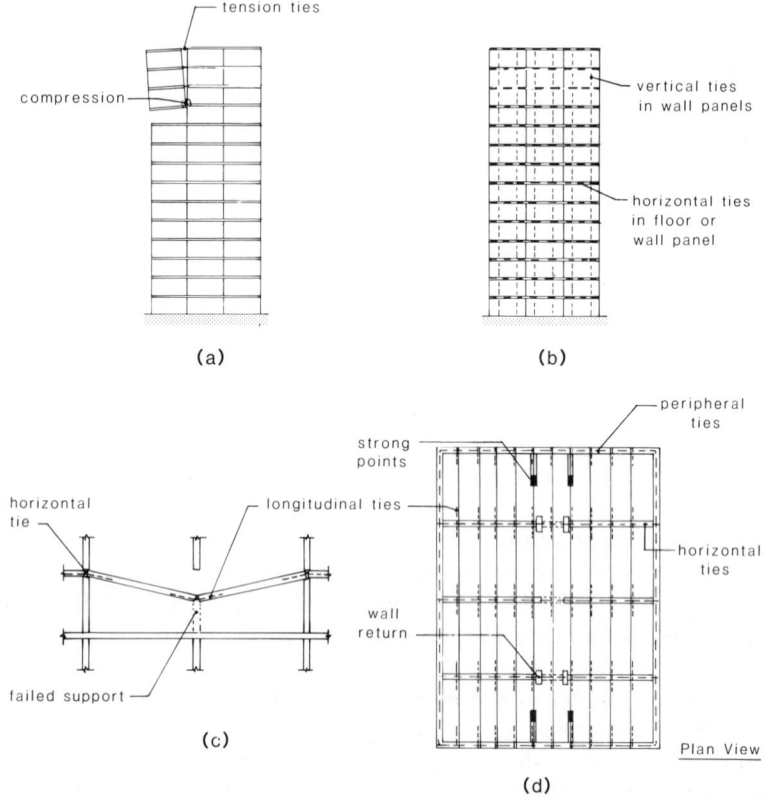

FIG. 9.3. Alternate load paths provided by tie reinforcement in precast concrete wall panel structures. (a) Alternate load path provided by horizontal and vertical ties; (b) horizontal and vertical ties; (c) preventing debris loading by longitudinal ties over supports; (d) horizontal, longitudinal and peripheral ties.

vertical ties provide the horizontal shear connection between the floor slabs and the wall panels, thus ensuring that the wall panels above the failed element act as a unit. These vertical ties must also be capable of transmitting a vertical force equal to the weight of the failed element. The horizontal and vertical ties are shown in Fig. 9.3(b). Fig. 9.3(c) illustrates the behaviour of a floor element after failure of one of the supports and also illustrates the roles of the longitudinal tie reinforcement placed in the direction of each span over the supports. These ties improve the ductility of the floor elements and provide post-failure resisting mechanisms which prevent serious impact loading on the floor below due to debris from the failing floor member. This resisting mechanism relies on the slip of the longitudinal tie in order to develop significant displacements and permit a suspension system to be formed. The Portland Cement Association (1977) prescribes a method of providing this kind of response using unstressed prestressing strand for these longitudinal ties.

Figure 9.3(d) shows the use of wall returns or strong points in wall elements in the form of reinforced columns with longitudinal steel and column tie reinforcement in order to provide additional stability in the event of a local failure. The continuous peripheral ties shown in Fig. 9.3(d) provide tension ties for floor diaphragm action and also provide a mechanism capable of resisting the loading if the corner support of the floor slab fails.

9.3 CAST-IN-PLACE CONCRETE STRUCTURES

9.3.1 Examples of Progressive Collapses

The 1973 Skyline Plaza apartment building collapse in Virginia which killed 14 workmen resulted in an investigation as reported by Leyendecker and Fattal (1973) and Carino *et al.* (1983). This dramatic collapse has been instrumental in making engineers aware that progressive collapse can also occur in cast-in-place slab construction. It is believed that a punching shear failure(s) occurred on the 23rd floor after the premature removal of forms supporting this floor and after a section of concrete had been placed on the 24th floor. Studies (Carino *et al.*, 1983) indicated that for the structure in its partially shored condition the punching shear capacity was exceeded in the sand-lightweight slab which was supported by rectangular columns having an aspect ratio as high as 4 to 1 (i.e. ratio of long side to short side).

Failure spread horizontally over an interior portion of the building under the area of newly placed concrete and spread vertically over the entire height of the building. The impact from debris and a construction crane falling on to the adjacent post-tensioned parking structure resulted in progression of failure both horizontally and vertically over this entire structure. Figure 9.4 shows the apartment tower and three-storey parking structure after collapse.

FIG. 9.4. Skyline Plaza Collapse. (Carino *et al.*, 1983 — Courtesy of *Concrete International: Design and Construction*.)

Feld (1964) reported on a 1956 failure of a flat plate office building in Michigan during construction. This collapse took place just after the placement of concrete on the fourth floor which was supported by the third floor which in turn was partially supported by reshores to the second floor. It is believed that failure was initiated by punching shear. It is noted that the slab contained service openings on two adjacent faces of some of the square columns. The failure spread horizontally and vertically leading to the total destruction of the east wing of the building.

The progressive collapse of a large portion of a 16-storey flat plate apartment building in Boston (Engineering News Record, 1971) resulted in the deaths of 4 workmen. It is believed that collapse was initiated by a punching shear failure around a rectangular column supporting the roof slab. At the time of the failure the floor slab of the mechanical penthouse was being placed above the area of initial failure.

The 1981 collapse of the Harbour Cay condominium in Florida resulted in the deaths of 11 workmen and led to an investigation as reported by Lew *et al.* (1982). Failure occurred during the casting of the roof slab of this five-storey flat plate concrete building. It is believed that the failure was initiated by punching shear in the slab at the fifth floor. It was concluded that the slab thickness was not sufficient for punching shear strength requirements and that the top reinforcement in the slab was placed about 1 in (25 mm) too low. The failure progressed horizontally and vertically resulting in the total collapse of the structure.

9.3.2 Susceptibility of Two-Way Slab Structures to Progressive Collapse

The four examples of collapse described above provide clear evidence of the susceptibility of flat plate structures to progressive collapse. It is important to realize that all of these failures occurred during construction and were initiated by punching shear failures. Although the four examples described were all flat plate structures it is conceivable that a punching shear failure in a flat slab structure could also trigger progressive collapse. There is no evidence to suggest that cast-in-place two-way slabs supported by beams between columns are particularly susceptible to progressive collapse.

9.3.3 Initiation of Failure in Flat Plate and Flat Slab Structures

Factors influencing initial failure in the form of punching shear are given below.

9.3.3.1 *Construction Loading Effects*

For flat plate and flat slab structures designed for low live loads the construction loading may be the most severe loading in the life of the structure. The construction loads cause large shears and unbalanced moments to be transferred through the slab–column connections when the concrete may not be fully cured. Agarwal and Gardner (1974) calculated a slab loading of $1 \cdot 17$–$2 \cdot 38$ of the self-weight of the slab to be considered during construction depending on the number of forms and reshores employed. Grundy and Kabaila (1963) observed that the maximum loads are always carried by the last slab cast before the shores at the lowest level are removed.

9.3.3.2 Design Factors Influencing Punching Shear Capacity

It is imperative that the influence of lightweight concrete, openings in slabs near columns, rectangularity of the columns and the effects of moment transfer be considered when checking the punching shear capacity as these have been contributing factors in past collapses. Figure 9.5 illustrates the moment–shear interaction relationship for a slab connected to an interior column. Loading in pure shear causes the limiting shear stress to be reached uniformly around the critical shear periphery. Loading along path OO' results in the limiting shear stress reached on face AB with zero shear stress on face CD. As suggested by Hawkins and Mitchell (1979), if the loading causes higher unbalanced moments to be transferred then an upwards shear stress occurs on face

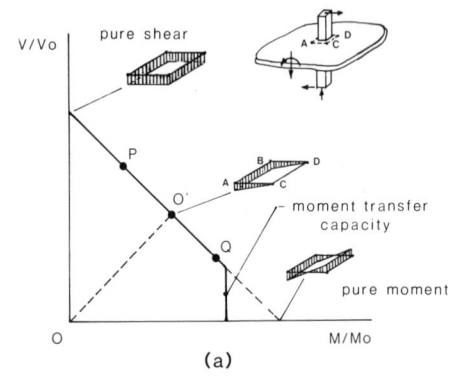

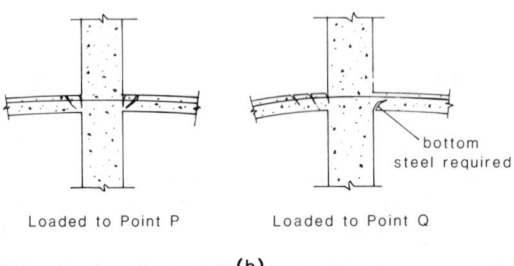

FIG. 9.5. Need for continuous bottom reinforcement in slab–column connections transferring large moments. (a) M-V interaction diagram for slab–column connection; (b) shear cracking in slab–column connection.

CD and may lead to cracking on the bottom of the slab as shown in Fig. 9.5(b). In these situations continuous bottom reinforcement would be required to prevent a premature tensile failure on the bottom surface of the slab and thus enable the expected moment–shear interaction capacity to be reached.

9.3.3.3 *Construction Errors*
Inadequate shoring or re-shoring was cited as contributing to two of the four collapses previously described. Improperly placed top reinforcement due to improper height of bar supports was cited as a contributing factor in the Florida condominium collapse. As reported by Lee *et al.* (1979) the punching shear capacity is particularly sensitive to changes in the effective depth of the reinforcement.

9.3.4 Progression of Failure in Flat Plate and Flat Slab Structures
In all four of the progressive collapses described the initial failure was triggered by punching shear failures around the columns. Figure 9.6

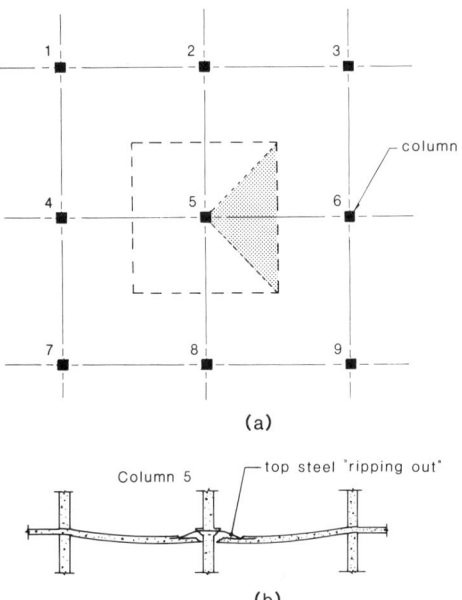

FIG. 9.6. Horizontal progression of failure in a flat plate structure not designed to prevent progressive collapse. (a) Plan view of flat plate; (b) elevation view during punching shear failure.

illustrates the brittle nature of a punching shear failure in a slab that has not been designed to prevent progressive collapse. After the sudden shear failure takes place at column 5, the top reinforcement rips-out of the top surface of the slab and becomes ineffective in carrying the load. This connection has no post-punching reserve strength and leads to abrupt loss of vertical support at the column location. Due to the loss of support at column 5 the load in the shaded area must be transferred in shear to column 6 and likewise shears are transferred to columns 2, 4 and 8. In addition the spans in both directions have doubled resulting in large increases in moments, thus requiring moment transfer through the slab column connections. This causes punching shear failures at columns 2, 4, 6 and 8, and leads to a horizontal progression of failure. A vertical progression of failure also occurs as the collapsing slab falls on to the floor below. Thus the chain reaction of failures has commenced with vertical progression of failure until the debris comes to rest at the base of the structure.

9.4 TENSILE MEMBRANE BEHAVIOUR OF TWO-WAY SLABS

The key in preventing progressive collapse of two-way slabs is to provide alternate load paths in the form of tensile membrane action.

9.4.1 Predicting Tensile Membrane Response

Figure 9.7 illustrates the idealized response for a two-way slab panel that is fully restrained at its edges and contains continuous two-way reinforcement that is well anchored at its edges. After the initial failure

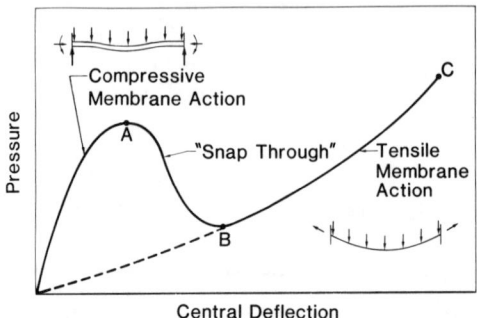

FIG. 9.7. Tensile membrane action.

has occurred, and after large deflections have taken place, tensile membrane action provides a secondary resisting mechanism which is capable of providing a significant resistance. Final collapse would occur after the steel reinforcement ruptures.

The three main analytical models for predicting tensile membrane response are described below.

9.4.1.1 Fully Plastic Membrane Equation

Park (1964) derived an equation describing the load versus central deflection membrane response of a uniformly loaded rectangular reinforced concrete panel with vertical and horizontal restraint at its edges as given below:

$$\frac{wl_x^2}{T_x\delta} = \frac{\pi^3}{4\sum_{n=1,3,\ldots}^{\infty} \left(\frac{1}{n^3}\right)(-1)^{0.5(n-1)} \left\{1 - \left[\cosh\left(\frac{n\pi l_y}{2l_x}\sqrt{\frac{T_x}{T_y}}\right)\right]^{-1}\right\}} \quad (9.1)$$

where w = uniformly distributed load per unit area of the panel; δ = central deflection corresponding to load w; l_x, l_y = clear span in the short and long directions respectively; T_x, T_y = yield force of the reinforcement per unit width, in the l_x and l_y directions respectively.

This equation assumes that all the reinforcement has yielded, that the reinforcement has a constant stress after yielding (fully plastic) and neglects any residual flexural resistance.

For an isotropically reinforced square panel the dimensionless parameter $wl_x^2/T_x\delta$ in eqn (9.1) becomes 13·56. Park suggested that the use of eqn (9.1) together with an assumed deflection equal to 0·10 times the shorter clear span would give a conservative estimate of the tensile membrane strength. Black (1975) suggested that a more accurate prediction of strength could be obtained by using Park's equation with a dimensionless parameter of 20·0 rather than 13·56 together with a limiting deflection of 0·15 times the shorter clear span.

9.4.1.2 Non-linear Iterative Method

Hawkins and Mitchell (1979) reported on a simplified iterative method for determining the tensile membrane response of panels having vertical and horizontal restraint at their edges. This method assumes that the membrane takes a circular deformed shape and neglects any residual flexural resistance. Equation (9.2) gives the equilibrium relationship for the circular membrane.

$$w = \frac{2T_x \sin \sqrt{6\varepsilon_x}}{l_x} + \frac{2T_y \sin \sqrt{6\varepsilon_y}}{l_y} \qquad (9.2)$$

where w = predicted load per unit area; l_x, l_y = clear span in the short and long directions respectively; T_x, T_y = force in the reinforcement per unit length in the x and y directions respectively, $\varepsilon_x, \varepsilon_y$ = strain in the x and y directions respectively.

For the assumed deflected shape the compatibility relationship between the central deflection δ, the geometry of the panel and the strain in the reinforcement is

$$\delta = \frac{3l_x \varepsilon_x}{2 \sin \sqrt{6\varepsilon_x}} \qquad (9.3)$$

Compatibility of deflections at midspan in the two directions gives the following relationship between ε_x and ε_y:

$$\varepsilon_y = \frac{\varepsilon_x l_x^2}{l_y^2} \qquad (9.4)$$

The complete load versus central deflection response can be obtained by using the following solution technique. (a) Choose a value of ε_x; (b) calculate ε_y from eqn (9.4); (c) determine T_x and T_y corresponding to ε_x and ε_y respectively, using the actual stress–strain relationship for the reinforcement; (d) calculate the load w using eqn (9.2); (e) calculate the deflection δ using eqn (9.3). The above steps give one point on the predicted load–deflection response. By repeating the steps for different values of ε_x, the complete tensile membrane response up to the rupture of the reinforcement can be predicted.

9.4.1.3 Computer Model Incorporating Non-linear Geometry and Non-linear Material Response

Mitchell and Cook (1984) reported on a computer model developed to predict the post-failure response of slabs. The slab reinforcement mesh is modelled by a grid of truss elements. In order to predict the response of the grid which experiences large deflections, the model accounts for the non-linearities arising from the changes in geometry (non-linear geometry) as well as the non-linearities arising from changes in element stiffnesses (non-linear material). The elements possess axial stiffness only and therefore any residual flexural resistance in the membrane is neglected. The computer model which utilizes the stiffness method of analysis takes account of the non-linearities by employing a step-wise iterative solution technique. The complete tensile membrane response can be predicted by solving for the displacements for a number of load increments. The advantage of this model is that the boundary conditions can be properly accounted for and no assumptions are needed concerning the deflected shape. This model provides a useful tool in investigating the tensile membrane response of single panels with edge restraint as well as the behaviour of typical two-way slab systems including flat plates and flat slabs which have more complex boundary conditions.

9.4.2 Experiments on Single Panels

Tests on single panel slabs to investigate tensile membrane response have been reported by Powell (1956), Wood (1961), Park (1964), Keenan (1969), Brotchie and Holley (1971), Black (1975) and Ritz *et al.* (1975).

Figure 9.8 illustrates the load versus central deflection responses of representative single panel slabs that were fully restrained around their edges. This figure also compares the measured responses with the responses predicted by Park's membrane equation, the non-linear iterative method and the non-linear computer model. Also shown is the deflection limit of $0.15\ l_n$ (where l_n is the clear span in the short direction) for each test. Table 9.1 gives the important parameters for each of these representative specimens.

As can be seen all three methods give fairly accurate predictions of the tensile membrane response of Brotchie and Holley's and Keenan's specimens. The conservative predictions for Park's and Black's specimens are probably due to the significant amount of top reinforcement around the perimeter of the fully restrained edges which resulted in considerable residual flexural resistance not accounted for in the predictions. Keenan's specimens contained continuous top reinforcement which was considered along with the bottom reinforcement in the predictions. The residual flexural resistance around the perimeter did not affect the membrane response of these large specimens. Due to difficulties with the loading systems the specimens tested by Park, Brotchie and Holley and Keenan were not loaded to final collapse. In spite of this difficulty four of the five responses shown in Fig. 9.8 reached deflections of about $0.15 l_n$.

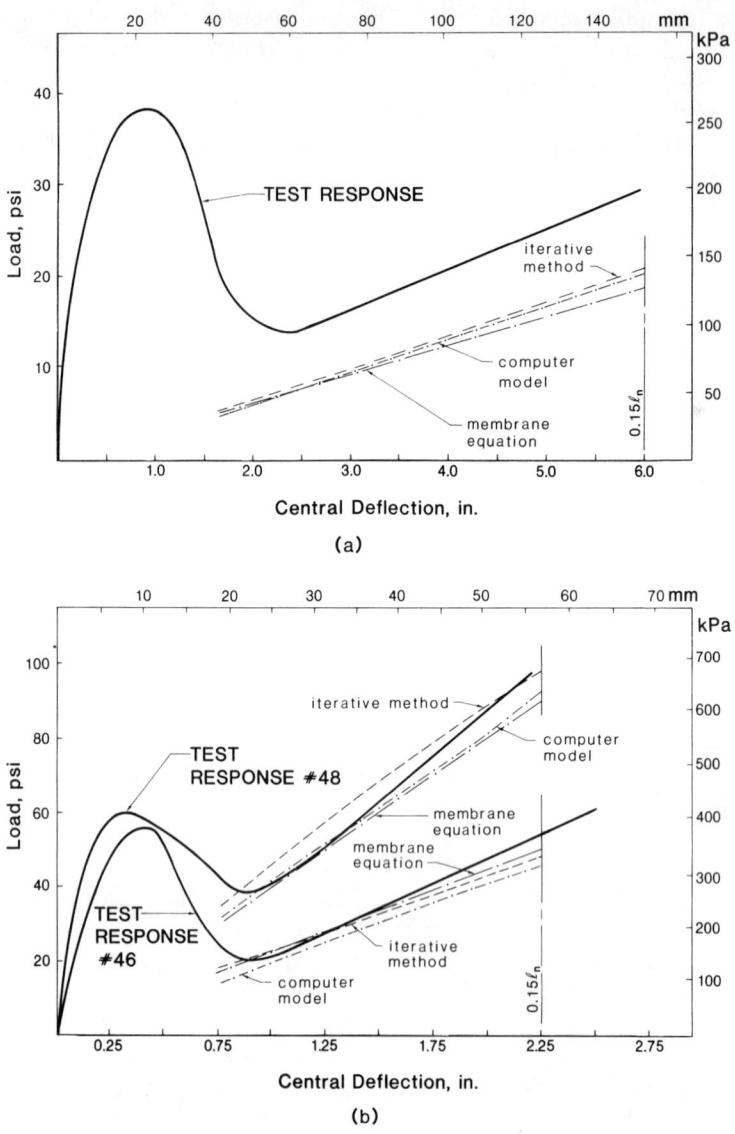

FIG. 9.8. Comparisons of predicted and observed responses for fully restrained single panel slabs tested by Park (1964), Brotchie and Holley (1971), Black (1975) and Keenan (1969). (a) Park's specimen A3; (b) Brotchie and Holley's specimens.

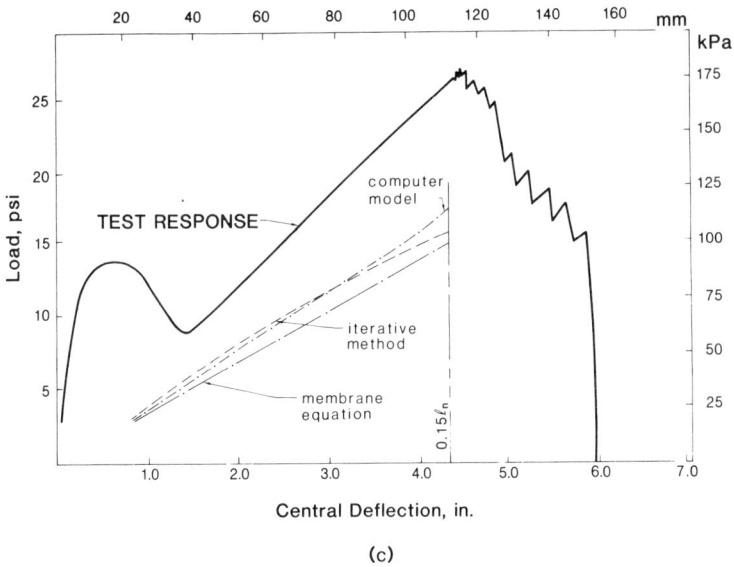

(c)

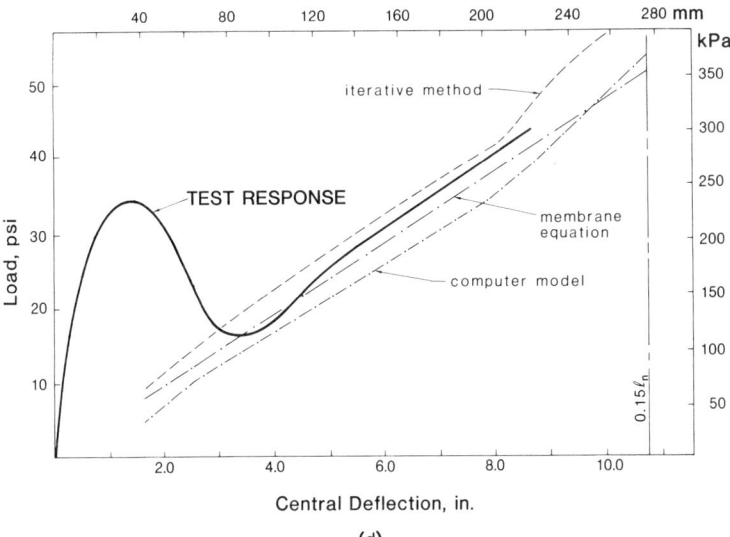

(d)

FIG. 9.8—contd. (c) Black's specimen IS3; (d) Keenan's specimen 3S3.

TABLE 9.1
PROPERTIES OF FULLY RESTRAINED SINGLE PANEL SLABS

		Dimensions			Bottom steel		Top steel			Comments
		l_x (mm)	l_y (mm)	t (mm)	ρ_x	ρ_y	ρ_x	ρ_y	f_y (MPa)	
Park (1964)	A3	1 016	1 524	50·8	0·590	0·161	1·18	1·18	352	Top steel not continuous
Brotchie & Holley (1971)	46	381	381	19·1	0·747	0·747	—	—	414	
	48	381	381	19·1	1·49	1·49	—	—	414	
Black (1975)	IS3	737	737	22·6	0·657	0·591	—	—	301	50% of bottom bars bent up at support
Keenan (1969)	3S3	1 829	1 829	76·2	0·611	0·611	0·611	0·611	342	Top steel fully continuous

A simply supported panel which does not have horizontal edge restraint has the ability to develop its own in-plane restraint around its perimeter provided that the edges are vertically supported and provided that the reinforcement is continuous and well anchored around the edges. The in-plane restraint is provided by the formation of a compression ring of concrete around the perimeter of the panel which equilibrates the tensions in the tensile membrane. This phenomenon is demonstrated by the results of tests carried out by Brotchie and Holley (1971). Figure 9.9 compares the load–deflection responses of the fully restrained panels (specimens 46 and 48) with the companion simply supported panels (specimens 34 and 35). Although the initial responses of similarly reinforced specimens are significantly different due to the boundary conditions, they are essentially the same after tensile membrane action develops. This effect has also been observed by Ritz *et al.* (1975) in the testing of prestressed slabs which had a variety of edge support conditions.

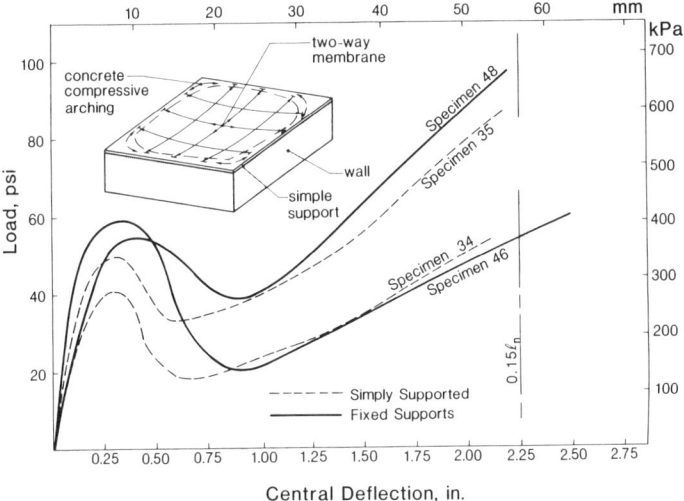

FIG. 9.9. Effect of edge restraint on response of single panel slabs tested by Brotchie and Holley (1971).

9.4.3 Experiments on Two-way Slab Structures

The development of tensile membranes as alternate load paths in two-way slab structures is described below.

9.4.3.1 *Two-way Slabs Supported on Stiff Beams or Walls*

A two-way slab panel supported on four sides by extremely stiff beams (or walls) behaves essentially the same as a single panel with vertical and horizontal edge restraints. For an interior panel the adjacent regions of the structure provide the in-plane horizontal restraint. A corner or edge panel develops its own in-plane horizontal restraint in the form of a compression ring.

The behaviour of a nine-panel slab structure supported on encased steel beams which was reported by Huff (1975) demonstrates the ability of a slab system with extremely stiff beams to develop tensile membrane action. This heavily reinforced structure was designed for blast resistance. Figure 9.10(a) shows the details of the slab structure which was loaded over its entire area. Figure 9.10(b) illustrates the significant tensile membrane response that was obtained in the central panel. The tensile membrane predictions, which are conservative, agree reasonably well with the measured response. The loading of the slab was discontinued due to rupturing of the reinforcing bars in a corner panel which contained a smaller amount of reinforcement than the central panel. The non-linear computer analysis and simplified iterative analysis give reasonably good predictions of the load corresponding to a deflection of $0 \cdot 15 l_n$. The fully plastic membrane equation gives more conservative results since no strain hardening of reinforcement is considered.

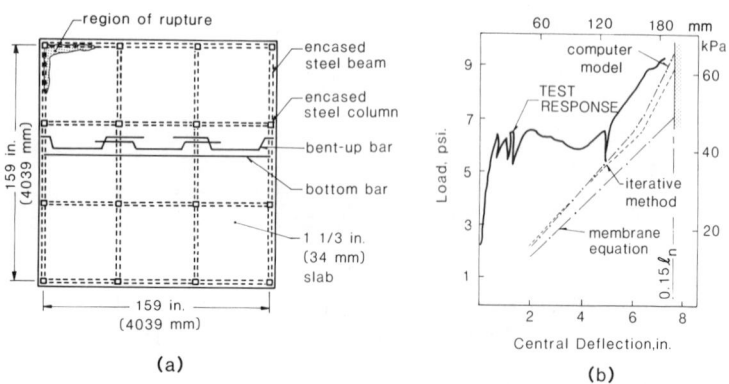

FIG. 9.10. Two-way slab structure supported on stiff beams tested by Huff (1975). (a) Details of slab structure; (b) predicted and measured responses of central panel.

9.4.3.2 *Flat Plate and Flat Slab Structures*

Since punching shear failures have played a prominent role in the initiation and progression of collapse, alternate load paths must be designed and detailed to provide post-failure resistance around slab-column regions. Figure 9.11 demonstrates the need for continuous bottom reinforcement properly anchored into the columns. Criswell (1972) and Hawkins and Mitchell (1979) reported that after shear failure has occurred the top reinforcement 'rips-out' of the top surface of the slab and becomes ineffective in carrying the load (see Fig. 9.11). A slab-column connection without bottom reinforcement properly anchored into the column would therefore have negligible post-punching shear resistance, which would result in the collapse of the slab and would likely cause progressive collapse of the structure (see Fig. 9.6). In contrast, bottom reinforcement well anchored into supports does not rip-out of the slab and thus provides significant dowel action, providing some post-punching shear resistance. If these bottom bars are both well anchored and effectively continuous then they will be capable of hanging the damaged slab off the columns as tensile membrane action develops (see Fig. 9.11). The severity of distress in this region emphasizes the need to carefully detail these bottom bars.

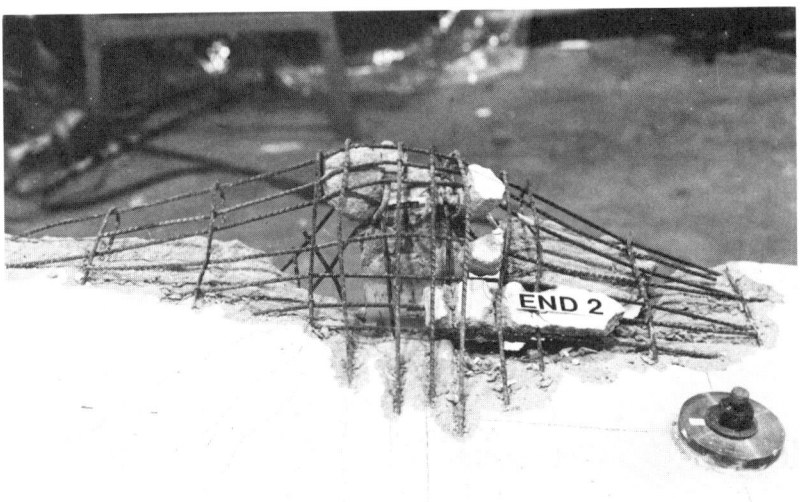

FIG. 9.11. Post-punching failure resistance showing 'ripping-out' of top steel and slab hanging by bottom bars (Chu, 1981).

Mitchell and Cook (1984) describe the alternate load paths shown in Fig. 9.12 which can develop in a typical two-way slab structure as a result of local failures. A well detailed interior panel would form a two-way membrane which is supported by one-way catenaries along column lines. These one-way catenaries consisting of continuous bottom reinforcement must be well anchored into the columns in order to transfer the load to the columns after punching shear failures have occurred.

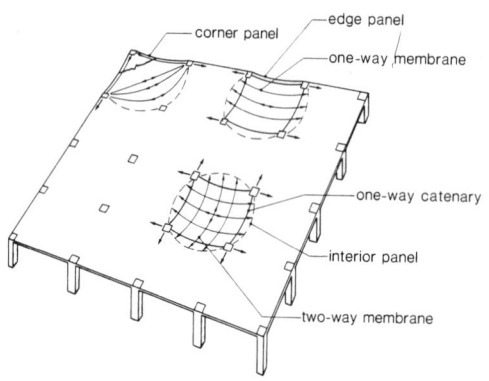

FIG. 9.12. Alternate load paths consisting of tensile membranes and one-way catenaries in two-way slab structures (Mitchell and Cook, 1984).

Figure 9.12 also shows the likely alternate load path for an edge panel. The edge panel forms a one-way membrane which spans between adjacent edge panels taking advantage of the available horizontal in-plane restraint provided by these panels. The one-way membrane is supported by one-way catenaries which span between the columns and run perpendicular to the free edge. The post-failure resisting mechanism for a corner panel (see Fig. 9.12) involves folding of the slab across the corner and the development of one-way catenaries diagonally across the corner of the slab seeking the horizontal in-plane restraint offered by adjacent panels.

Figure 9.13 shows the details of the nine-panel flat slab structure tested by Criswell (1972). The structure contained both drop panels and column capitals and was subjected to uniform load by means of water pressure. Tension lap splices were used to provide continuity of the top and bottom reinforcement in the slab which was designed for blast loading. However, at some drop panel edges all the top and bottom reinforcement was lap spliced. These lap splices performed well up to and beyond the initial failure load as tensile membrane action developed. Deflections of at least $0 \cdot 18 l_n$ were measured in the one-way catenaries between a column and the perimeter wall, with failure occurring by failure of the lap splices.

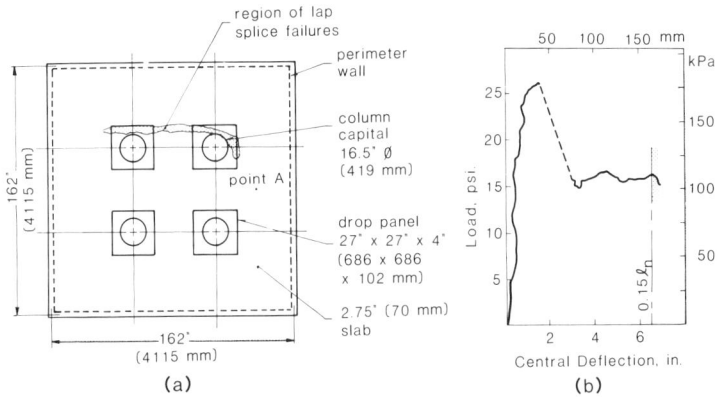

FIG. 9.13. Flat slab structure tested by Criswell (1972). (a) Details of slab structure; (b) load vs deflection response at point A.

Figure 9.14 describes the experimental set-up and experimental load deflection responses of flat plate corner panels reported by Chu (1981) and Balazic (1982). The one-quarter scale test structures had overloads placed on one corner panel at a time with simulated uniform loading. All three specimens had identical spans of 1500 mm (59 in), 100 mm (3·94 in) square columns and slab thicknesses of 47·5 mm (1·87 in) and similar concrete material properties. The slabs were designed and detailed in accordance with the ACI Standard 318-77 (1977) for the same design load.

Corner panels CP1 and CP2 contained two D2 and four D2 continuous bottom bars respectively (area of one D2 is 12·9 mm² (0·02 in²)), which were anchored into the columns. As can be seen in Fig. 9.14 both these panels, which had no beams, exhibited final collapse loads which exceeded the initial failure loads. The additional anchored bottom bars in panel CP2 resulted in a larger initial failure load and a larger collapse load than was obtained in panel CP1. Corner panel CP3 had 100 mm (3·94 in) square spandrel beams with three D4 bottom bars (area of one D4 is 25·8 mm² (0·04 in²)) well-anchored into the columns. The beneficial effects of strengthening and stiffening the free edges of a corner panel are evident if the responses given in Fig. 9.14 are compared.

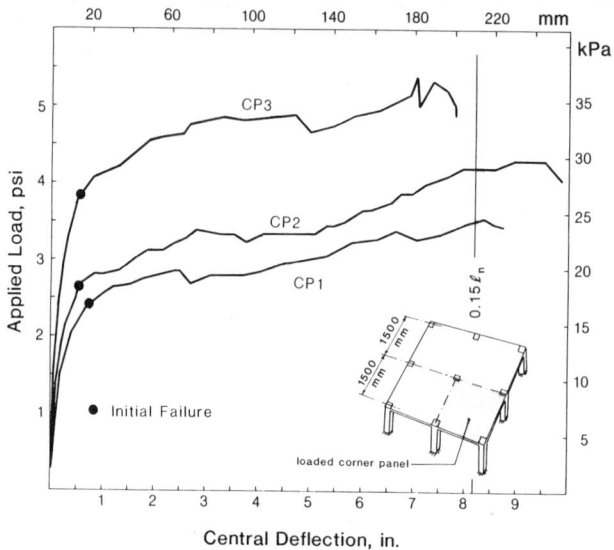

FIG. 9.14. Flat plate corner panels tested by Chu (1981) and Balazic (1982).

The corner panels responded as described in Fig. 9.12 with the slab folding across the corner with the continuous well anchored bottom bars transferring the load to the columns.

9.4.3.3 *Typical Two-way Slab Supported on Beams*
Hopkins and Park (1971) reported the results of testing a nine-panel two-way slab supported on beams. The slab structure had continuous bottom reinforcement in the slab and the beams contained stirrups. This experiment illustrated the ability of such a structure to fully develop tensile membrane action.

9.5 DESIGN OF TWO-WAY SLABS FOR ALTERNATE LOAD PATHS

9.5.1 Design Philosophy
The philosophy of design is to provide alternate load paths in the form of two-way tensile membranes and one-way catenaries in order to

prevent progression of failure. Current design and detailing practice for two-way slabs provides sufficient two-way reinforcement in the positive moment regions of slab panels in order to develop the required two-way tensile membrane.

The most important aspect of the design is to provide continuous bottom reinforcement between the columns and ensure that this reinforcement is well anchored into the columns. These continuous bottom bars serve the following functions:

(a) prevent initial punching shear failure due to large unbalanced moments to be transferred through the slab–column connections (see Fig. 9.5);
(b) provide the tension in the one-way catenaries between columns and ensure the transfer of tensile forces into adjacent slab panels (see Figs 9.11 and 9.12).

Mitchell and Cook (1984) developed a design expression for the amount of reinforcement in the one-way catenary by assuming that the one-way catenary is subjected to a uniform load equal to $0 \cdot 5wl_2$ per unit length. This arises from the vertical components of the forces in the bars of the two-way membrane which is hanging off the one-way catenary. Thus from eqn (9.2) the one-way catenary response can be described as

$$w = \frac{4T_c \sin\sqrt{6\varepsilon}}{l_n l_2} \quad (9.5)$$

where T_c = force in the anchored bottom reinforcement in the one-way catenary;
l_n = clear span of the one-way catenary;
l_2 = distance measured from the centreline of the panel on one side of the catenary to the centreline of the panel on the other side of the catenary.

If a limiting deflection of $0 \cdot 15 l_n$ is chosen for the one-way catenary then from eqn (9.3) the strain in the reinforcement is $0 \cdot 0538$ and eqn (9.5) becomes

$$T_c = 0 \cdot 465 \, w l_n l_2 \quad (9.6)$$

The prediction from eqn (9.6) is compared with test results for the nine-panel structures tested by Criswell (1972) shown in Fig. 9.13. In the region where the tension splices failed the total area of bottom bars that passed over the effective column capital region was $0 \cdot 44$ in^2 (285 mm^2).

These bars have a yield stress of 36·85 ksi (254 MPa) thus giving a catenary force, at yield, equal to 16·21 kips (72·1 kN). The clear span between the edge of the column capital and the edge of the perimeter wall is 43 in (1092 mm) and l_2 equals 54 in (1372 mm).

Thus the failure load of the catenary from eqn (9.6) is 15 psi (103·5 kPa) for this slab structure which was loaded over its entire area. This compares well with the actual load measured on the slab at a deflection of $0·15l_n$ (see Fig. 9.13b).

9.5.2 Design and Detailing Recommendations

The design and detailing recommendations developed by Mitchell and Cook (1984) have been incorporated in the Canadian Standards Association code (1984) and are described below.

The area of continuous bottom steel A_{sb} required in the direction l_n is

$$A_{sb} = \frac{0·5 w_s l_n l_2}{\phi_s f_y} \qquad (9.7)$$

where A_{sb} = minimum area of effectively continuous bottom bars in direction l_n placed through reaction area of supports;
w_s = load to be carried after initial failure, assumed to be the larger of the total service load acting on the slab or twice the self-weight of the slab;
l_n = clear span in the direction being considered, measured face-to-face of supports;
l_2 = distance measured from the centreline of the panel on one side of the catenary to the centreline of the panel on the other side of the catenary;
f_y = specified yield strength of non-prestressed reinforcement;
ϕ_s = material resistance factor for reinforcing bars = 0·85.

The bottom bars having an area of A_{sb} may be considered effectively continuous if (a) they are lap spliced within a support reaction area with reinforcement in adjacent spans with a minimum lap length of l_d, or (b) they are lap spliced immediately outside of a support reaction area with a minimum lap splice of $2l_d$ provided the lap splice occurs within a region containing top reinforcement, or (c) they are bent, hooked or otherwise anchored at discontinuous edges to develop the yield stress at the support face. When the calculated values of A_{sb} differ for adjacent spans (due to unequal spans or unequal loading) then the larger value of A_{sb} must be used in each of these spans to enable the transfer of tension across the supporting member.

The unit loading w_s in eqn (9.7) should not be taken less than the total unfactored service loading for the intended use of the structure. However, since most progressive collapses have occurred during construction it is suggested that the loading not be taken less than twice the slab dead load which corresponds to typical load levels during construction as reported by Agarwal and Gardner (1974). It is emphasized that these design recommendations are not intended to prevent progressive collapses of typical two-way slab structures that are grossly overloaded over a large portion of the structure.

In post-tensioned two-way slabs the area of draped prestressing steel passing through the columns or reaction areas may be assumed effective. The term $\phi_s f_y$ is replaced with the term $\phi_p f_{ps}$ in eqn (9.7).

Figure 9.15 illustrates the special details of bottom bars of area A_{sb} (shown for one direction only) for a flat plate structure and a flat slab structure. It is noted that the design recommendations typically do not

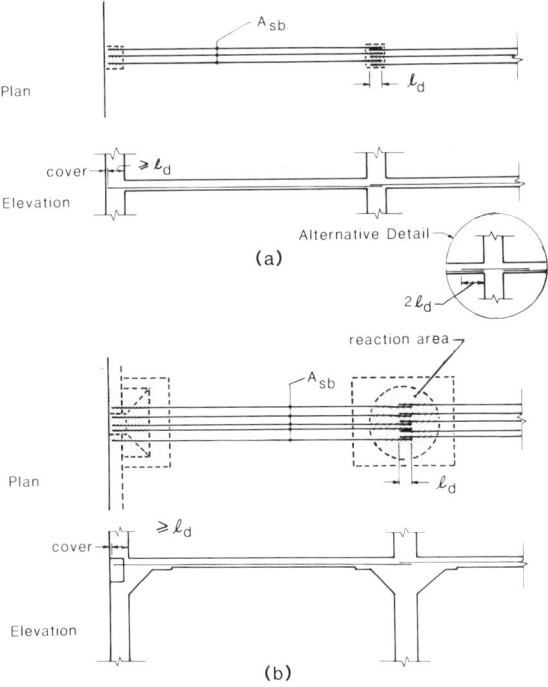

FIG. 9.15. Bottom bars designed and detailed to provide alternate load paths (bottom bars shown in one direction only). (a) Flat plate structure; (b) flat slab with drop panels and column capitals.

require additional amounts of bottom steel from existing design practice. Instead these requirements result in a rearrangement of the bottom steel such that an area A_{sb} of bottom bars pass through the column regions in each span direction and these bottom bars are detailed such that they are effectively continuous.

REFERENCES

AGARWAL, R. K. and GARDNER, N. J. (1974) Form and shore requirements for multistory flat slab type buildings. *Journal of the American Concrete Institute,* **71** (11), 559–69.

AMERICAN CONCRETE INSTITUTE COMMITTEE 318 (1977) *Building Code Requirements for Reinforced Concrete,* ACI 318-77, American Concrete Institute, Detroit, 102 pp.

BALAZIC, J. M. (1982) Tests to Collapse of Concrete Slabs. M.Eng. Thesis, McGill University, Montreal, Canada, 125 pp.

BLACK, M. S. (1975) Ultimate strength study of two-way concrete slabs. *Journal of the Structural Division, ASCE,* **101** (ST1), 311–24.

BROTCHIE, J. F. and HOLLEY, M. J. (1971) *Membrane Action in Slabs,* SP-30, American Concrete Institute, Detroit, pp. 345–77.

BURNETT, E. F. P. (1975) Abnormal loading and building safety. In *Industrialization in Concrete Building Construction,* SP-48, American Concrete Institute, Detroit, pp. 141–75.

BURNETT, E. F. P., SOMES, N. F. and LEYENDECKER, E. V. (1973) *Residential Buildings and Gas-related Explosions,* NBSIR 73-208, Center for Building Technology, National Bureau of Standards, Washington.

CANADIAN STANDARDS ASSOCIATION (1984) *Design of Concrete Structures for Buildings,* CAN3-A23.3-M84, 281 pp.

CARINO, N. J., WOODWARD, K. A., LEYENDECKER, E. V. and FATTAL, G. S. (1983) A review of the Skyline Plaza collapse. *Concrete International: Design and Construction,* **5** (July), 35–42.

CHU, E. K. (1981) Tests to Collapse of Reinforced Concrete Flat Plates. M.Eng. Thesis, McGill University, Montreal, Canada, 108 pp.

CRISWELL, M. E. (1972) *Design and Testing of a Blast-resistant Reinforced Concrete Slab System,* Technical Report No. N-72-10, Weapons Effects Laboratory, US Army Engineering Waterways Experimental Station, Vicksburg, Miss., 312 pp.

ENGINEERING NEWS RECORD (1971) Building collapse blamed on design. *Construction,* **187** (3), 15 July, 119.

FELD, J. (1964) *Lessons from Failures in Concrete Structures,* Monograph No. 1, American Concrete Institute Monograph Series, pp. 30–2.

GRIFFITHS, H., PUGSLEY, A. and SAUNDERS, O. (1968) *Report of the Inquiry into the Collapse of Flats at Ronan Point, Canning Town,* Ministry of Housing and Local Government, HMSO, London, 71 pp.

GRUNDY, P. and KABAILA, A. (1963) Construction loads on slabs with shored formwork in multistory buildings. *American Concrete Institute Journal, Proceedings,* **60** (12), 1729–38.

HAWKINS, N. and MITCHELL, D. (1979) Progressive collapse of flat plate structures. *American Concrete Institute Journal,* Technical Paper Title No. 76-36, *Proceedings,* **76** (7), 755–808.

HOPKINS, D. C. and PARK, R. J. (1971) Tests on a reinforced slab and beam floor designed with allowance for membrane action. In *Cracking, Deflection, and Ultimate Load of Concrete Slab Systems,* SP-30, American Concrete Institute, Detroit, pp. 223–50.

HUFF, W. L. (1975) *Collapse Strength of a Two-Way Reinforced Concrete Slab Contained within a Steel Frame Structure,* Technical Report No. N-75-2, Weapons Effects Laboratory, US Army Engineering Waterways Experimental Station, Vicksburg, Miss., 204 pp.

KEENAN, W. A. (1969) *Strength and Behaviour of Restrained Reinforced Concrete Slabs under Static and Dynamic Loadings,* Technical Report R621, US Naval Civil Engineering Laboratory, Port Hueneme, Calif., 131 pp.

LEE, Y. M., MITCHELL, D. and HARRIS, P. J. (1979) Lessons from structural performance — slabs containing improperly placed reinforcing. *Concrete International: Design and Construction,* **1** (6), 45–53.

LEW, H. S., CARINO, N. J. and FATTAL, S. G. (1982) Cause of the condominium collapse in Cocoa Beach, Florida. *Concrete International: Design and Construction,* **4** (8), 64–73.

LEYENDECKER, E. V. and FATTAL, S. G. (1973) *Investigation of the Skyline Plaza Collapse in Fairfax County, Virginia,* Report BSS 94, Centre for Building Technology, Institute for Applied Technology, National Bureau of Standards, Washington, D.C., 57 pp.

MINISTRY OF HOUSING AND LOCAL GOVERNMENT (1970) *The Building (Fifth Amendment) Regulations,* HMSO, London.

MITCHELL, D. and COOK, W. D. (1984) Preventing progressive collapse of slab structures. *Journal of the Structural Division, ASCE,* **110** (7), 1513–32.

PARK, R. (1964) Tensile membrane behaviour of uniformly loaded rectangular reinforced concrete slabs with fully restrained edges. *Magazine of Concrete Research, London,* **16** (46), 39–44.

PORTLAND CEMENT ASSOCIATION (1976) *Design and Construction of Large-Panel Concrete Structures — Report 2 — Philosophy of Structural Response to Normal and Abnormal Loads,* PCA, Skokie, Illinois, August, 133 pp.

PORTLAND CEMENT ASSOCIATION (1977) *Design and Construction of Large-Panel Concrete Structures — Report 4 — A Design Approach to General Structural Integrity,* PCA, Skokie, Illinois, October, 156 pp.

POWELL, D. S. (1956) The Ultimate Strength of Concrete Panels Subjected to Uniformly Distributed Loads. M.Sc. Thesis, Cambridge University at Cambridge, UK.

RITZ, P., MARTI, P. and THÜRLIMANN, B. (1975) *Versuche über das Biegeverhalten von vorgespannten Platten ohne Verbund,* Institut für Baustatik und Konstruktion Zürich, Zürich, Schweiz, Juni, 144 pp.

WOOD, R. H. (1961) *Plastic and Elastic Design of Slabs and Plates, with Particular Reference to Reinforced Concrete Floor Slabs,* Ronald Press Co., New York, 344 pp.

INDEX

Abnormal loadings, 250-2
ACI Code, 12, 18, 20, 21, 22, 25, 36, 170, 171, 178, 229, 231, 232, 233, 239, 242, 243

Beam
  group, design constraints, 102
  sections, optimal design, 97-9
Bending
  biaxial, 67, 111-32
  moments, elastically restrained column, 8
  torsion and shear combined, and, 158-63
  torsion combined, and, 152-8
Braced frames, 7-8, 25
British Code CP110, 13, 15, 21
British Steel Structures Code, 36
BS 8110, 231, 238-41, 245
Buckling, lateral, 67

Canadian Code, 25
Cast-in-place concrete structures, progressive collapses in, 255-60
CEB-FIP Model Code, 164, 170, 171, 178, 238, 239, 242, 243

CEB Model
  Code, 12-13, 15, 19, 22
  Column method, 15, 22
Chimneys, 58-9
CIRIA
  Guide, 170, 171, 175, 178-86, 188
  Report, 229
  supplementary rules, 181-3
Colombian Code, 26, 36
Column capitals, 218
Column group
  design constraint, 102
  objective function for, 101-2
Column moments
  imposed deformations, due to, 33-5
  imposed loads, due to, 33-5
Column sections, optimal design, 99-101
Columns
  braced frames, 7-8, 25
    modification of magnifiers to apply to, 19-21
  partially braced frames, 25
    critical loads, 35-6
  unbraced frames, 8-10, 25
    critical loads, 35-6
  combined axial load and biaxial bending, subjected to, 111-32
  different end moments, with, 63-4

Columns—*contd.*
  frames, in, 7–10
  isolated, 63–4
  see also Hinged columns;
        Nonslender columns;
        Slab-column connections;
        Slender columns
Column-supported shear walls, 193–216
  analysis of, 197–210
    approximate approach, 197–200
    comparison of design approaches, 206–10
    experimental approaches, 206
    finite element method, 206
    frame idealization, 200–2
  description of the problem, 195–7
  gravity loading, 195–6
  lateral loads, 196–7
  remarks concerning behaviour, analysis and design, 210–14
Compatibility compression field theory, 164
Compression field theory, 163
Compression members, stability problem, 43–69
Computer(s)
  programming, 89
  software
    design optimisation, 104
    nonlinear analysis, 103–4
    practical applications, and, 103–4
Cracked element, 95
Creep
  deflections, 18–19
  eccentricity procedure, 19
Critical load factor, 36
Cyclic loading, 76

Deep beams, 169–91
  bearing capacity, 181, 183
  crack control, 181
  definition, 170
  design codes, 178–86
  design example, 183–6

Deep beams—*contd.*
  elastic behaviour, 171
  failure modes, 172–5
  shear
    capacity, 180, 182
    span/depth ratio, 172–5
    strength, 175–8
  strength in bending, 179, 181
  ultimate load behaviour, 171–5
  web openings, 181, 186–8
Deformation
  analysis, 76
  characteristics, 78–82
Design Member Linking Strategy, 101
Design optimisation, 96–103
  computer software, 104
DFP (Davidon-Fletcher-Powell) algorithm, 103
Direct design method, 64–5
Direct Search Method, 103
Draft Australian Code, 17
Drop panels, 218

Eccentricity-curvature diagram for slender columns, 15
Equivalent frame methods for flat slabs, 224–37
Euler buckling load, 11

Failure
  criteria, 94
  initiation,
    flat plate and flat slab structures, in, 257–9
    progression in precast concrete buildings, 252–3
  mechanisms, 222–3
  modes, 140, 172–5
  slab structures, in, 259–60
  surfaces, 140
  theories, 147
Finite element analysis, 74, 83, 92–5, 206
FIP recommendations on practical design, 18

Flat slabs, 217–48
  basic equilibrium statement, 220
  behaviour of, 219–24
  crack patterns, 222, 223
  deformation due to moment from column, 226
  equilibrium requirement, 219
  equivalent frame methods, 224–37
  failure mechanisms, 222–3
  local rotational displacements, 221
  local vertical displacements, 221
  overall deformation on columns, 221
  punching
    failure modes, 240–2
    shear, 237–46
  serviceability in relation to deflections, 246
  shear reinforcement, 240–2
  transverse distributions of moments, 232–7
  see also Slab-column connections
Flexural rigidity, 80–2

General Limit Method, 86–8
Geometrical imperfections, 55

Hinged columns, 3–6
  moment magnifier for, 11–13
  sustained loads, effect of, 6
  unequal end moments, with, 5–6
Hinged system, 85
Hooke's law, 171

Imposed Rotations Method of Macchi, 74
Inelastic rotation capacity, 82
Influence diagrams, 86, 87

Kinnunen/Nylander punching model, 238

Lateral buckling, 67

Lateral drift effects, 9, 23–32
  calculation of, 26–32
  EI values, 30
  out-of-plumb construction, 31
  resistance factors, 31
  stage of loading, 30
Limit state
  collapse, of, 73, 74, 82, 99
  serviceability, of, 73, 74
Limit State Design, 73
Linear programming model, 96–7
Load-moment history in braced frame, 8

Masts, 60
Material
  characteristics, 75–7
  failures, 5, 17
Mathematical programming
  model, 89–92
  techniques, 74
Member stability effect, 23–32
Moment
  diagrams, 27
  magnification, 3–5, 124–7
  magnifiers,
    hinged columns, for, 11–13
    modification of, 19–21
    modification of elastically derived, 13–19
Moment–Curvature relationship, 14, 57, 78–9, 82
Moment-rotation characteristics, 80, 81, 83, 88, 91
Multi-axial stress state, 77

Nonlinear analysis, 82–95
  computer software, 103–4
Nonlinear programming
  formulation, 97–103
  model, 101
Nonslender columns
  combined axial load and biaxial bending, under, 114–20
  design methods, 116–20
  load-deformation response, 116

INDEX

Nonslender columns—*contd.*
   stiffness of, 120

P-Δ-method, 63
Parity rule, 88, 89
Plastic rotation capacity, 83
Plasticity compression field theory, 163
Poles, 60
Precast concrete
   buildings, failure initiation and progression in, 252–3
   structures, 252–5
     alternate load paths, 254
Precast wall panel structures, structural integrity of, 253–5
Progressive collapse
   cast-in-place concrete structures, in, 255–60
   definition of, 250
   philosophy of preventing, 252
   slab structures, of, 249–77
   two-way slab structures, of, 257
Punching
   failure modes, 240–2
   shear, 237–46, 257, 258, 269

Reinforced concrete frames, optimal design, 101–3
Ronan Point collapse, 251, 252
Rotation
   adjustment, 88–9
   capacity of hinging regions, 82

Safety factors, 50–1
St. Venant theory, 135
Sequential Unconstrained Minimisation Technique, 103
Shear
   bending and torsion combined, and, 158–63
   retention factor, 95
   walls supported on columns. *See* Column-supported shear walls

Short columns. *See* Nonslender columns
Simplified Limit Method, 74, 83–9
Skyline Plaza collapse, 255, 256
Slab–column connections
   continuous bottom reinforcement, 258, 269
   corner columns, with, 245–6
   eccentric loads, with, 242
   edge columns, with, 243–5
   M-V interaction diagram, 258
   models of, 229
   moment-shear interaction, 244
   shear cracking in, 258
   shear reinforcement, 240–2
   stiffness of, 225–30
Slab structures
   construction loading effects, 257
   experiments on single panels, 263–72
   failure in, 259–60
   fully restrained single panel, 266
   progressive collapse of, 249–77
   tensile membrane behaviour, 260–72
   response investigation, 263–72
   two-way, 257
     alternate load paths, 270, 272–6
     experiments on, 267–72
     supported on beams, 272
     supported on stiff beams or walls, 268
     tensile membrane behaviour, 260–72
Slender columns
   analysis procedure, 121–3
   combined axial load and biaxial bending, under, 120–8
   correction for unequal end moments, 13
   design procedures, 123–8
     1950 to 1975; 10–23
     recent developments, 23–35
   design shortcomings, 22–3
   eccentricity–curvature diagram for, 15
   load, under, 4, 5
   moment–curvature diagram for, 14

Slender columns—*contd.*
  second-order analysis, 127–8
  unbraced frames, in, 126
Slender reinforced concrete
    members, comparison
    of design methods, 47
Slender walls, 67–8
Slenderness bounds, 21–22
Slenderness effect, 5, 6, 114
Slenderness ratio, 124
Stability
  check, 49–58
  factor (Q), 36
  failures, 5, 16
  reinforced concrete building
    frames, of, 1–41
Steel reinforcement, 92–3, 95
Stiffness
  matrix, 89, 90, 93–4
  nonslender columns, of, 120
Storey magnifier method, 27
Strain energy method, 139
Stress–strain relationships, 51–4,
    75–8, 94, 95, 115
Structural imperfections, 55
Structural integrity of precast wall
    panel structures, 253–5
Sustained loads, 6, 18–19
Sway frame equilibrium, 10
Sway frames, 21, 61–3
Swiss code design procedure,
    12

Tensile membrane response
  computer model, 262–3
  fully plastic membrane equation,
    261

Tensile membrane response—*contd.*
  non-linear iterative method, 261–2
  prediction of, 260–7
Tension
  failures, 17
  stiffening effects, 53–5
Torque, expressions for, 142–3, 145–6
Torsion, 133–68
  bending and shear combined,
    and, 158–63
  bending combined, and, 152–8
  design recommendations, 163–5
  plain concrete members, 135–8
  prestressed concrete beams,
    146–52
  recent developments, 163–5
  reinforced concrete in, 138–46
Torsional buckling load, 38
Torsional cracking of plain concrete
    beam, 139
Torsional reinforcement, 145
Torsional stability of three-
    dimensional buildings,
    37–8
Torsional strength, 145

Ultimate load analysis, 88
Uniaxial stress state, 75

Variable stiffness method, 32–3

Waffle slabs, 218, 234

Yield criteria, 94

RAYMOND H. FOGLER LIBRARY
DATE DUE

BOOKS ARE SUBJECT TO RECALL AFTER TWO WEEKS

MAR 27 1987

NOV 14 1987